T0343783

HOW TO ARGUE WITH A **MEAT EATER**

Nelson Mandela once stated that, 'Education is the most powerful weapon which you can use to change the world.' This book is dedicated to that very idea.

Nelson Mandela, Speech, Madison Park High School, Boston, 23 June 1990

HOW TO ARGUE WITH A MEAT EATER

MEAT EATER

EATER

(AND WIN EVERY TIME)

ED WINTERS

Vermilion
LONDON

1

Vermilion, an imprint of Ebury Publishing
One Embassy Gardens, 8 Viaduct Gdns,
Nine Elms, London SW11 7BW

Vermilion is part of the Penguin Random House group of companies
whose addresses can be found at global.penguinrandomhouse.com

Penguin
Random House
UK

Copyright © Ed Winters 2024

Ed Winters has asserted his right to be identified as the author of this Work in
accordance with the Copyright, Designs and Patents Act 1988

No part of this book may be used or reproduced in any manner for the purpose
of training artificial intelligence technologies or systems. In accordance with
Article 4(3) of the DSM Directive 2019/790, Penguin Random House
expressly reserves this work from the text and data mining exception.

First published by Vermilion in 2023

This edition published by Vermilion in 2024

www.penguin.co.uk

A CIP catalogue record for this book is available from the British Library

ISBN 9781785044496

Printed and bound in Great Britain by Clays Ltd, Elcograf S.p.A.

The authorised representative in the EEA is Penguin Random House Ireland,
Morrison Chambers, 32 Nassau Street, Dublin D02 YH68

Penguin Random House is committed to a sustainable future
for our business, our readers and our planet. This book is
made from Forest Stewardship Council® certified paper.

AUTHOR'S NOTE

When I first went vegan, I took it upon myself to learn as much as I possibly could about veganism. I read books, watched documentaries, listened to academics and delved into the scientific literature. Throughout this process I also began to learn more about the opposing sides of the argument, yet the more I delved into the arguments of those who disagreed with veganism, the more confident and reassured of my own position I became. It didn't take long before I then started advocating vocally and publicly for veganism.

Throughout my years advocating for veganism I have encountered every argument against veganism that exists, from the expected to the eyebrow raising. One of the main reasons for this is because I have put myself in countless situations where I have actively invited and encouraged people to give me their arguments against veganism.

Growing up, the prospect of actively seeking out diehard cowboys in the heartlands of Texas was never high on my list of priorities. However, after deciding I wanted to have interesting conversations about veganism with people, all of a sudden chatting with diehard cowboys became an altogether more intriguing idea.

Now, many years after I initially started talking about veganism, I've had the privilege of discussing it face-to-face in front of thousands of different people. These experiences have not only been illuminating in terms of understanding and hearing all the different arguments against veganism that exist, but have also been invaluable in allowing me to stress test my rebuttals and responses. These conversations have essentially presented me with an abundance of opportunities to become a better communicator and a more accomplished advocate.

At the same time, I've also been immersed in the ever-growing body of scientific literature that discusses the issues related to our use of animals. I find myself constantly in awe of the incredible work and dedication being carried out by academics, data analysts and scientists, while also being dismayed by the misinformation and disinformation that sadly still spreads like wildfire.

I have also had the great honour to meet so many wonderful vegans who have shared their experiences with me and whose stories have inspired me deeply. From people who have been vegan for decades, to people who have just made the change – each person's experience being individual and unique, yet often with difficulties and hardships that overlapped.

Sadly, time after time the vegans I meet tell me that they struggle to effectively talk about veganism with the people in their life, or they say they feel overwhelmed by all the arguments people use.

How to Argue With a Meat Eater is a book that has allowed me to bring together all of these different aspects of my work, drawing from all of these experiences and from everything that I have learnt and have had the privilege to be a part of so far. It is because of all of this that this book has become so important and personal to me. It is my hope now that this book becomes of great value to you too.

CONTENTS

Introduction   1

Understanding People's Arguments   7
Becoming a More Effective Debater:
   The Core Skills You Need   27
The Arguments   44
The Mistaken Philosopher   45
The Egotist   56
The Wishful Thinker   63
The Social Conformer   77
The 'What-Abouter'   85
The Compromiser   99
The Anti-Woke Warrior   116
The Well-Intentioned Leftist   130
The Misinformed Environmentalist   146
The Animal Farmer   170
The Pseudoscientist   191
The Naturalist   223
The Historian   239
The Amateur Nutritionist   250
The Practicalist   287

Afterword   297
Notes   303
Acknowledgements   327
Index   329
About the Author   339

INTRODUCTION

Whether or not you're vegan, the chances are you've had some form of engagement with someone about veganism. Maybe with a friend or family member, maybe with someone online. There's even a strong chance that that interaction resulted in an argument or a heated discussion, such is the provocative nature of veganism.

Perhaps even just by reading the word 'veganism' your mind has begun to think of the arguments you have heard or even perhaps used yourself to dismiss it. After all, aren't we told that eating animals is part of the food chain, animal products taste nice and almonds are responsible for killing bees and using loads of water?

The truth is, veganism can mean many different things to different people. Depending on who you ask, veganism can be anything from colonialist, racist, privileged, extreme and dogmatic to another symptom of the woke snowflake culture stripping people of their freedoms. Not to mention it might even be tricking people into unknowingly cannibalising (yes, really).

So, it's certainly not hyperbolic to say that there is a plethora of arguments that people use against veganism, and that there are a lot of questions and accusations that often get bandied about, even when they lack substance. It is for this reason, as a vocal advocate for veganism, that I never find myself short of work, continually rolling the rock up the mountain before something new is levelled at me and the rock rolls back down again.

It's also why I wrote this book. I know all too well the challenges that vegans face and how difficult it is to counter the many different arguments that are levelled against us. However, by laying out these arguments and equipping you with the ability to confidently defend yourself, it is my hope that, unlike for Sisyphus, one day the rock won't roll back down again.

THE ARCHETYPES

Cannibalism aside, many of the arguments against veganism don't seem unreasonable at first and are actually understandable when you realise how little most people know about how we treat animals and the issues that veganism covers. I also believe that it is important for ideas and beliefs to be challenged, and their rationales to be scrutinised, even if only to stress-test why we believe what we do. Veganism is no exception.

However, while I welcome logical, genuine and well-intentioned questioning of veganism, people often fall into the trap of parroting arguments or fallacies that they've read or heard elsewhere without fact-checking them. It is all too easy to see something online that aligns with what we want or believe to be true and never explore whether it actually is.

It is for this reason that this book will do two things. First, the opening chapters will arm you with the knowledge to become a more effective debater by exploring why people use the arguments they use, and also by discussing tips and techniques that will help you to debate more effectively. This will allow you to feel more confident and be more impactful in your conversations, ensuring that we can all have a healthier and more constructive dialogue around an issue that can often fall into unproductive disputes and tribalism.

Second, the book will outline the main arguments against veganism and provide a response to them. Because the claims you hear often depend on who is putting them forward, I have grouped them by the archetypes I have come across during my years of advocating for veganism. While these groupings can be a bit tongue-in-cheek, they showcase the wide variety of arguments that are levelled against vegans, as well as the people who present them.

The information contained within this book comes not only from extensive academic and scientific research but also from my years of experience debating people from all walks of life and with myriad different beliefs. Throughout the book I have included snippets from and references to many of these conversations and debates, including with hunters, farmers, TV presenters, heads of farming unions and lobby groups, and, of course, everyday meat eaters. I have included these to illustrate real-world examples of the arguments and rebuttals, and to provide further context to the topics being discussed.

In essence, this book is a vegan's best friend. It's the one I wish I had been able to read when I first stopped eating animal products and began advocating for veganism. If you are a vegan, it will leave you feeling empowered to walk into your next family gathering or your workplace knowing that you will not only be able to face but also rebut any argument that is thrown your way.

However, this book is not only for vegans. It is also for any non-vegan who is interested in scrutinising their beliefs and who is open to exploring the rebuttals to the arguments levelled against vegans. In fact, for those of you who are not vegan, I hope that this book presents you with the opportunity and intellectual stimulation to look at your lifestyle choices in a way that you have never done before and, in doing so, it inspires you to recognise the power you have to create a positive difference every single day. After all, this book is a tool for empowering ourselves to become more critical thinkers, better conversationalists, and more informed, educated and motivated citizens – something that is important for all of us, vegan and non-vegan alike.

LESSONS FROM 427 BC

In 427 BC, a debate took place in Athens regarding the fate of the Mytileneans, a group of people who had revolted against Athenian rule. During his ultimately successful bid to convince the assembly to not slaughter all the Mytilenean men and enslave all the women and children, a Greek citizen called Diodotus said, 'The good citizen ought to triumph not by frightening his opponents but by beating them fairly in argument.'

Unfortunately, many of the most important issues we face are played out through the weaponisation of fear. Every election seems to be a testament to that. The same is true for how veganism is discussed in the media, with Piers Morgan once claiming that vegan protestors were 'terrorising old women', and also describing himself as the 'vegan resistance', as if he sees himself as the Luke Skywalker of the anti-woke vegan rebellion, armed not with a lightsaber but a footlong meat sausage roll and diminishing arterial health.

Pro-meat hysteria is not just isolated to the UK. In 2021, Fox News hosts had a meltdown. Biden's climate plans had outraged them because they were going to lead to red meat consumption being slashed by 90 per cent. Except there was a slight problem with their reporting: it wasn't true.

Much to my disappointment, there was no plan to encourage Americans to reduce their consumption of red meat, let alone by 90 per cent. But who needs a fair argument when you can frighten people instead? Unfortunately, nearly 2,500 years later, Diodotus could still make the same point.

The reason I mention these examples is because I strongly believe that the discourse around veganism is often reductive, ill-informed and insincere. Rather than being critically reflected on and analysed through a lens of curiosity, it is instead another victim of the transgressive desire to manufacture outrage at the cost of societal progression.

However, in my time advocating for veganism, I have been fortunate enough to speak to thousands of people about this very issue, and what has filled me with optimism and a constantly renewed sense of purpose is that, when given the opportunity and space to thoughtfully engage with the topic of veganism, people are not only understanding of its merits but also will often find themselves agreeing with its principles more than they imagined they would. And for good reason.

Food is powerful and empowering. After all, how often do we have a choice in front of us that has consequences for the most pressing issues that our species face? Environmental collapse, antibiotic resistance, pandemic-causing infectious diseases, and, of course, the biggest driver of animal exploitation and suffering caused by humans. Our food choices present us with an opportunity to take a stand against all of these issues and many more.

And that is another reason why this book exists. I wrote it because our food choices are of the utmost importance, and because I want you to feel inspired, informed and empowered.

So use this book to arm yourself with the facts you need. Read it and re-read it so that you can respond to any argument that your family member, friend or colleague puts to you. Use the citations in the back of the book to look at all the research I draw on throughout, and feel emboldened to know that you can inspire others to reflect on their own choices and make important changes. This book is your tool, so use it as such. Annotate it, underline points that are especially important to you, and make it your own.

Most importantly, I hope that by reading this book you can see that when given the opportunity to be fairly argued, veganism not only holds up to scrutiny but is also, as we vegans often claim, essential if we are to create a more ethical, kind and sustainable world.

UNDERSTANDING PEOPLE'S ARGUMENTS

At a talk I did in a school, a student asked me, 'Do you dislike everyone who isn't vegan?'

On the one hand I was saddened by the question, but I also understood why the pupil had asked it. I had spent the previous 40 minutes giving a presentation on the merits of veganism using words like 'immoral' and 'wrong' to describe the consumption of animal products. The obvious inference from this type of language is that I view people who are not vegan as immoral people, and the idea that I would dislike them was therefore understandable.

However, by no means do I dislike everyone who isn't vegan. I have friends and family members who aren't vegan, and I've had many incredible experiences in the company of non-vegans as well.

I might strongly believe that what we do to animals is immoral but context matters. We live in a complex and highly nuanced world where uncomfortable truths are hidden from us. Where behaviours, practices and industries that cause harm are normalised and ubiquitous. Where our laws and the behaviours of those around us reinforce the behaviours that we engage in.

Polarising issues are divisive for a reason, and the very nature of their polarisation can make it hard to ascertain which position is the one we should support.

When it comes to veganism, if you add in misinformation, disinformation and propaganda, and sprinkle on some psychological and sociological behavioural drivers, then it is no surprise that a belief system that is still supported by only a small percentage of the population is challenged and resisted by so many people.

Does that mean that all non-vegans are bad people who shouldn't be liked? I believe not. Throughout my years of discussing and debating veganism I have learned many valuable lessons. One of the most important being that if we don't understand why people think or act the way they do, it becomes much harder to change their minds about something. Now, understanding someone isn't the same as agreeing with them. We can understand why someone thinks the way they do and still fundamentally disagree with, oppose and challenge them.

However, attempting to understand why someone thinks, says and does the things they do is an incredibly important aspect of creating constructive and effective conversations. This is because it allows us to empathise with their position and rationalise why they believe the things that they do. This in turn means that when we are responding to them we can do so from a place of nuance that allows us to address their arguments in the best way possible.

For example, I often speak to international university students. If one were to say to me that they ate certain animal products because those products were prominent in their culture, while I know that this argument doesn't hold up to scrutiny (see 'The Social Conformer'), I might surmise that they were in part making it because they were currently living in a different culture to the one they grew up in, with all of the challenges that this presents. Perhaps they feel isolated or homesick as a result.

Because of this rationalisation, I would then respond in a

mindful way while at the same time highlighting that something isn't morally justified just because it's a cultural norm. In the next chapter ('Becoming a More Effective Debater: The Core Skills You Need'), I discuss specific techniques and tips for responding to situations like this in an effective way.

NOW YOU KNOW. AND KNOWING
IS HALF THE BATTLE

As a species, we are sometimes the victims of our own success and are a little complacent due to our intelligence. Undeniably, we are capable of some truly incredible things but being human also brings with it imperfections, flaws, and everything else that comes as a consequence of our complexity.

The interconnectedness of our world and the growing coverage of so many complex and global issues has created a situation where we feel the need to be informed about more things than ever. We feel we have to formulate opinions on topics we know very little about, decide what our beliefs are about issues that scholars and academics dedicate their whole careers to trying to understand, and come to terms with scientific concepts that are normally investigated by experts.

Today we don't just worry about food, shelter and companionship, but also about an event that happened thousands of miles away, or even something that is invisible to the naked eye or existential in nature.

These more abstract concepts, which often exist outside of our own immediate environments, are especially susceptible to unsubstantiated and erroneous beliefs and assumptions. While it is much harder to be convinced that up is down when up and down are right in front of us, if up and down are out of sight and contradict what we want to believe, it becomes a much easier sell to convince us of the opposite.

Social media has created the perfect environment for complex ideas to be hijacked by those who know how to appeal to the masses and do so in a way that turns them against the scientific consensus. Who needs science when a social-media influencer is telling me up is down and down is up in 60 seconds? Oh, and I better make sure to subscribe to their monthly academy where they promise they'll also tell me that left is right.

The G.I. Joe fallacy – from the *G.I. Joe* cartoon series in which each episode ended with the closing tagline 'Now you know. And knowing is half the battle' – is the incorrect idea that simply being aware of a bias is enough to overcome it. We might think, *Well, I know that people spread misinformation and that lobby groups shouldn't be trusted, so I won't be fooled.* But that knowledge doesn't mean we are immune to the biases we know exist.

Unfortunately, knowing is not half the battle. This is why being critical of the information we receive is also important, as is taking the time to investigate claims before simply accepting them. It can be useful to scrutinise why something is the way it is, or why we might be being told something. Does this piece of information reinforce or contradict what we've been told previously and how does it fit into what we want to be true?

For example, we might be aware of how biases exist but then see a post about someone who says they almost died because they went vegan, all their teeth fell out, their skin went grey, they had less protein than a sea cucumber and their brain was foggier than a coastal town on a cold morning. All of a sudden that unsubstantiated anecdote creates an emotional reaction that reading through a scientific paper about the health benefits of a plant-based diet never could.

However, if we can gain a sense of perspective by acknowledging the response we've had and why we've had it, we can gain some sense of objectivity and further explore the claims being made to see if they're substantiated.

Anecdotes more often than not provide the weakest evidence but are frequently the most effective means of dictating what people actually believe about something. And the thing is, most of us know this to be true, yet anecdotes and personal stories still remain one of the best ways to convince someone of something, irrespective of what the evidence actually shows.

Because of faulty anecdotal evidence, you can find people who will swear that drinking urine cured them of every ailment and that drinking bleach is the cancer-curing medicine that the pharmaceutical companies want to keep hidden from us. These people place too much emphasis on the opinions and anecdotes of non-authority figures and in turn ignore the work of experts, academics, scientists and organisations whose life work and research should form the basis of what they understand to be true.

Even from the perspective of veganism, I often get asked during interviews about how going vegan made me feel, but that's the wrong question. The question should be: what does the body of evidence show?

In practical terms, one of the most impactful questions I ask people is 'How do you know that?', because asking people how they know something is a great way of exposing biases or problems with their sources of information. One of the most telling responses to that question I had was when talking to a fisherman who stated that fish don't feel pain. I asked him how he knew, and he replied, 'From other fishermen.' In another conversation, a student stated in regard to animals, 'They don't have feelings in the same way we do.' In response I asked him, 'How do you know that?' To which he smiled sheepishly and said, 'I don't know that.'

Exposing people's biases and the flaws in their sources of information increases self-awareness, as it forces them to scrutinise their beliefs and can convince them to reconsider their positions. Importantly, it also encourages them to become more critical in their decision-making and of their sources moving forward. Sadly, misinformation and disinformation spread so widely because of a lack of evaluation from those receiving it, so revealing this fallibility to people can improve their information literacy.

CONVINCING OURSELVES OF WHAT IT IS WE DESIRE

In 1912, a man named Charles Dawson claimed that he had made a discovery in East Sussex of hugely important significance. He had uncovered the fossilised remains of a previously unknown species of early human that became known as the Piltdown Man. It was claimed that this discovery was the so-called 'missing link' between ape and man.

In the nineteenth century, archaeologists in France had discovered Cro-Magnons and archaeologists in Germany had discovered the Neanderthals. Just five years earlier than Dawson's discovery, Heidelberg Man had been discovered in Germany as well. However, Britain had no such findings to its name – until Piltdown Man. European and American scientists approached Piltdown Man with a considerably larger amount of scepticism than their British counterparts, suggesting that there had been a mistake, as the findings fundamentally altered our understanding of human evolution. However, in the UK, the story was for the most part accepted in good faith.

Then, in 1953, it was finally confirmed what had been

suspected by some all along: Piltdown Man wasn't real. It had been a fraudulent hoax carried out by a scammer and an opportunist, and improperly challenged by a nation desperate to make its archaeological imprint and reinforce nationalistic pride during a time of increasing conflict and competitiveness in Europe.

Unfortunately, when influenced by strong and persuasive drivers, people's ability to make rational decisions can be impacted by their desire for something to be true. This is called 'motivated reasoning'. Simply put, people use their emotional biases to create justifications or make decisions rather than being guided by what is accurately shown by the evidence. This is pertinent to many areas of life, including what people do to animals.

What meat eater wouldn't want to believe that farmed animals are good for the environment and that eating them isn't an ethical concern? All of a sudden they would be able to eat steak and not feel like they were betraying their desire for climate action. They would be able to have what they wanted and still think of themselves as being ethical at the same time. What people want to be true often becomes what they convince themselves is true.

People are constantly searching for justifications that not only allow them to keep eating animal products but mini-mise or completely invalidate the reasons why they shouldn't. Does it matter if the evidence doesn't support that position? Well, that all depends on how badly they want to keep eating steak.

Motivated reasoning is similar to confirmation bias, whereby people have a tendency to seek out information that aligns with their existing beliefs or what they desire to be true. They then ignore information that is inconsistent with their beliefs, opting to simply reaffirm what they want to be true, rather than investigating what is actually true.

Case in point: a post went viral on social media in early 2023 in which a photo of the bones belonging to a vegan person was depicted alongside one of the bones belonging to a non-vegan. The vegan bones were shown to be less strong and less dense, and the caption accompanying the images stated that plant-based diets supposedly increase the risk of bone fractures. The post was viewed and reshared by millions of people. There was a problem, though. The photos weren't comparing vegan bones and non-vegan bones at all – they were images taken by NASA to show the impact that being in space has on bone density.

The person who originally shared the misrepresented photos believes that people who don't eat red meat are doing so because of mental weakness and have been programmed by nefarious groups that want people to stop eating meat so that they can be controlled. This person is the Charles Dawson and meat eaters are the eager audience too excited and ready to believe what he has to say to actually stop and verify it. And this is why confirmation bias is so dangerous. Who's going to spend time reverse-image-searching and fact-checking the information they are seeing on social media, especially when what they're seeing is actually what they want to be true? After all, if you are desperate for Piltdown Man to be real, you'll convince yourself he is.

Understanding how motivated reasoning works is essential, as it shows the importance of using good-quality sources. If we are advocating for something, being aware of what trusted and highly regarded sources are publishing is an important way of arming ourselves with the necessary information. It also means we are more capable of identifying less credible information or claims that contradict the scientific evidence.

So be on the lookout for credible scientific journals, studies that analyse data from a multitude of different studies (referred to as meta-analyses), and a consistent theme within the scientific literature. If you read something in a newspaper article, in a blog or on social media, try and find the root of the claims to see whether what you have read has been substantiated.

If you ask someone how they know something and they can't remember off the top of their head where they saw the information, ask them to send over their sources after the conversation has ended. Likewise, offer to share your sources with them. That way they can have a read through in their own time and juxtapose their claims and sources with yours.

I THINK THEREFORE I'M RIGHT

People often overestimate their own competence or knowledge about a certain issue. I've lost count of the number of times people have confidently asserted to me that it is vegans who are destroying the rainforests because of their soya consumption, or that humans don't do things to animals that are in fact commonplace. This type of bias is known as the Dunning–Kruger effect, in which people without knowledge in a certain area feel as if they have the authority to make assertions about it.

Veganism is one of those issues that very few people have actually explored properly yet most have an opinion or belief about. Instead of seeking a substantiated view, people perceive themselves as knowing more about it than they actually do. Because veganism pertains to our individual choices and their moral impact, it's a topic that people especially want to view themselves as being right about.

Importantly, because there is an increasing amount of conversation about veganism, what people do to animals is now often discussed in terms relating to ethics and sustainability. This puts consumers in a tricky situation. Being ignorant about something that people are being told is important presents them with a potentially uncomfortable situation. They can fall into the trap of wilful ignorance, but while this might provide somewhat of a safety blanket from any cognitive dissonance, ultimately trying to remain ignorant about veganism is revealing in itself.

Alternatively, the need to validate and defend the consumption of animals means people simply align with the information that tells them what they want to hear. That way they can feel like their actions are substantiated and they are more educated about the issues than they actually are. This is problematic, as not only does it disincentivise people from researching further, but it can give them a reinvigorated sense of conviction around what they're doing.

This is why it's important that people take time to reflect on their beliefs, which is an especially pertinent point for any non-vegans reading this, but applies to all of us more generally too. We should question ourselves and where the information we are consuming is coming from. It's important that we get out of the echo chambers that we can find ourselves in. Perhaps we can talk to someone who disagrees with our position and listen to their viewpoint and how they challenge ours. Or perhaps we can simply research the opposing point of view to our own.

While it can be frustrating, recognising that people often have a limited understanding of the issues can also push us to be more empathetic and patient. After all, even some vegans had negative opinions and beliefs about veganism before they made the change. So this understanding makes us more respectful and constructive. Rather than being condescending towards someone, we can use our own experiences and awareness of how people can change to instead foster a growth mindset and promote intellectual honesty through a critical reflection of their beliefs.

Rather than being exasperated by people's lack of understanding, we can use the opportunity to focus on education and learning so that we can point out the gaps in their knowledge and hopefully empower them to take the time afterwards to actually look into the issues with an open mind.

PERSPECTIVE BEYOND POLITICS

As veganism continues to get caught up in the culture wars, the way that people view veganism is increasingly susceptible to the bias of their political identities, meaning that they form opinions based on their political stances rather than on the merits of veganism and the evidence for and against it.

The danger of such thinking is it creates tribalism and pigeonholes something in such a way as to make it alienating to huge swathes of people. It is dangerous to feel that you have to think a certain way, or you should look for a certain type of criticism of something simply based on how you wish to be perceived and identify.

There is also the fear of being alienated from a community or diluting a form of identity that people hold to be important. Does a so-called freedom-loving, Fox News-watching Republican hate veganism because they have evidence-based reasons to hate it or because they think they should? What would going vegan mean for their identity credentials? If they told their freedom-loving, Fox News-watching Republican friends that they no longer ate steak, would that make them seem a little bit less freedom-loving, or even a little bit less of a patriot?

So, at what point are our views truly a reflection of what

we believe as opposed to merely a product or extension of what we think we should think or want to think based on our environment, our past experiences, our peers, or our desire to uphold a certain identity?

The clothes we wear allow us to showcase our identities and helps others understand what our personalities are like. They allow us to present ourselves to the world in the manner in which we wish to be presented. Our food choices and moral beliefs can also be an extension of how we want others to perceive us. However, when we view something based on how it lines up with how we wish to be perceived we diminish ourselves as critical thinkers.

Of course, what I have just described is perfectly fine when people's outward-facing appearances and beliefs are informed by things such as musical taste. However, when they are dealing with issues that concern the rights and wellbeing of others, to adopt this approach means they are no longer acting from a place of rationality, something that is incredibly important when dealing with matters of this serious nature.

Discussing issues rationally using evidence and critical thinking creates a more collaborative approach and allows people to make informed choices. After all, these are not just decisions that pertain to an individual's life and experiences – thinking rationally allows people to encompass others who are also impacted by the choices they make. This is especially important considering non-human animals are unable to represent themselves in the same way that we can. Not to mention they are also kept away, locked in farms and slaughterhouses, where people's ability to empathise with them is reduced. Because of this, it is vital that people view this as an issue that exists beyond the end of their own noses. To do so requires them to be intellectually honest and reflective.

Unfortunately, people often operate using their ego, viewing the world from the position of 'I' and making statements like 'I don't want to change' or 'I want to eat meat.' In doing so, they are framing the issue as being one whose importance is superseded by what they want, rather than what is conducive to a better world all round.

People's detachment from what happens to animals means it's hardly surprising that for so many the question of veganism is grounded in the question of whether or not they want to stop eating animals, rather than whether stopping eating animals is the right thing to do. So, in such a situation, it is wise to find commonality and understanding, rather than attacking the person for an argument that appears selfish in nature.

One way to do this is to simply explore the edges of a person's argument so that you are creating a rebuttal within their moral framework. An example of this came up during a debate I had with a student called David who made the argument that it's acceptable to eat animals because it's pleasurable.

Me: So it's OK to do something to an animal if we can get pleasure from it?

David: If it's painless.

Me: So what if we tranquillised an animal and then sexually exploited them?

David: No, I don't agree.

Me: No? Why not though?

David: I don't know. I would just not feel comfortable.

Me: They're not suffering and it's for enjoyment.

David: That's true. That's a valid point.

Me: So why is that wrong?

David: I don't know. It's a good question.

Instead of making a statement about why killing animals because they taste good is wrong, I instead used his argument to illustrate that not even he subscribes to it when it's consistently applied.

Sadly, viewing veganism through the lens of political identity, culture war issues or just from the position of each individual means that the animals are overlooked, as are all the other problems associated with animal products. This means we need to encourage people to view these issues from a different perspective. We can do this by prompting them to reflect on how they would feel if they were in a similar situation to an animal. Alternatively, perhaps they have a pet and we can ask them to think about how they would feel if someone treated them in the same way we treat other animals.

We can also appeal to shared values: do they think we should care for animals? Are they worried about the climate crisis? We can highlight the interconnectedness of these issues and how the consequences of animal farming impact them too. And as well as exploring the edges of their arguments, we can ask people to explain their political beliefs if it's relevant to the conversation. If they're a fiscal conservative, how do they feel about the huge financial handouts given by the State to prop up animal farmers? If they're a socialist, what do they think about the monopolisation and corruption found within the animal farming industries, and what about social justice with regard to animals?

SHOOT THE MESSENGER

Vegans are against farmers. Vegans are against meat eaters. Vegans are against freedom and personal choice. Vegans are against normal behaviours and are extreme. People consistently view vegans as attempting to create conflict and being inherently antagonistic. In doing so it can create the impression that being vegan is not desirable.

One of the easiest ways to not engage with someone is to perceive that person as being dislikeable. The same works with veganism. If people can paint an image of veganism as being something that is wholly unlikeable, or can think of vegans as conforming to an undesirable expectation, they can instantly distance themselves from what it is we are advocating for.

Now, don't get me wrong, vegans can have an optics problem. We can sometimes be our own worst enemies, presenting a version of veganism or a characterisation of what being a vegan means that does significantly more harm than good. The stereotype of the aggressive activist shouting at people springs to mind. Yet descending into ad-hominem attacks or ascribing certain traits to vegans to make veganism seem unappealing is a disingenuous way of attempting to avoid the arguments we make.

However, it does present us vegans with an opportunity to leave someone with a more positive impression of veganism. This is why, as vegans, one of the most important things we can do to soften peoples' opinion of veganism is to not conform to the stereotype that they might expect from us. If the person we are talking to expects that we will shout at them or be unreasonable, one of the main positives of not doing so is that we will lead that person to doubt their initial negative perception and create a more positive impression of veganism in general. Plus, by removing one of the main barriers that people use in order to not engage with the issues, we can create a situation where that person is more likely to discuss what is truly important: the merits of and arguments for veganism.

Vegans are often the ones labelled as irrational or overly emotional. And while I would concede that caring about the suffering of others has an emotional aspect to it, caring about the rights of others is also a deeply logical position to take. It is not irrational to challenge the inconsistencies and contradictions inherent in people's treatment of some animals compared to others, nor is it a sign of emotional weakness to be appalled by the infliction of harm on sentient beings.

So while vegans are accused of acting from a place of irrationality, the irony is that even when presented with the rebuttals to their arguments, many non-vegans still won't change their minds and will continue to support industries that are responsible for doing things that violate many of the beliefs that they say they have. After all, who isn't against animal cruelty? More or less everybody says that they are, but are people actually against animal cruelty when they continue to support the industries responsible for causing more human-inflicted suffering, harm and death to animals than any other?

Rationally speaking we would surely say not. Yet here we are in a situation in which the same people who say they are against animal cruelty are in fact supporting and perpetuating the industries responsible for the worst examples of it. Try as hard as they might, that square piece will just not fit in the round hole.

The fact that some people don't change their views when presented with rational rebuttals of their arguments might sound rather demoralising. However, it is well documented that people's minds are changed through repetition. So it's not surprising that many people don't necessarily switch their points of view the first time they are presented with important information, or have their worldview and beliefs challenged. We are creatures of habit who find comfort in the status quo and, as such, to change someone's behaviours often requires planting seeds and encouraging them to germinate.

ALIGNING OUR MORAL SENSE OF SELF

This is why it's so important to understand the strong connection between beliefs, actions and psychology, because clearly it's not as simple as knowing that something is an issue, or indeed even knowing what the facts are. People are complex and so are the drivers of our behaviours.

People's consumption of animal products is tied to so many other aspects of their lives, from their desire to be comfortable and not isolate themselves from cultural norms or cause friction in their personal lives to how they identify as moral beings with agency. Eating food in particular is not simply about the acquisition of calories and nutrients. Food is an integral part of people's families, social dynamics and cultural identities. A person might have fond memories of eating a roast dinner with their family, or perhaps their grandma makes the best lasagne and they have shared so many happy experiences with her because of it. Perhaps they have always gone to the same restaurant with their partner or ordered the same takeaway with their friends. Food is pleasure, it is social, it is a mechanism through which experiences can occur and memories can be created. Food is a tool through which we can feel prosperous, connected and a part of something bigger than us.

To understate all that food represents and all that it has historically represented is to do a disservice to something that is of the utmost importance if we are ever to create healthier conversations around food choices.

When we talk about veganism, we are not simply discussing the merits of eating plants over animals – we are fundamentally challenging all of these different aspects of people's lives and identities. Can that person share those experiences with their grandma again once they've made the decision that killing animals for lasagne is immoral? What impact will no longer being able to eat with their partner at the same restaurant mean for the memories they've created over foods they now consider to be ethically abhorrent? What does viewing eating animal products as unethical mean for their perception of the kind of person they are? Do they want to think of themselves as being the causers of harm to others? And do they want to confront the immorality of their previous actions and reconcile the dissonance and conflict that creates within them?

I wish the arguments against veganism were valid. I wish that eating animals were more ethical, more sustainable and healthier than not eating them. In the same way, I wish that fast fashion were great for the workers and for the environment. I wish that there were no such thing as the climate crisis. I wish that all of these things weren't issues, because who wouldn't? How much easier would it be if our food choices weren't a problem? How much simpler would it be if we could just eat the way that was the most convenient for us and for that to also be the most positive food choice?

I so desperately wish that the climate science around animal consumption and being morally consistent when it comes to our treatment of animals were wrong, because the consequence of being right is the suffocating reality that there is a gargantuan problem that the vast majority of people desperately try to convince themselves isn't a problem at all. However, try as hard as they might to believe that it is acceptable to consume animals, the issue remains regardless.

Refusing to accept the existence of something does not mean that thing does not exist. People must instead confront the daunting prospect of change versus the daunting prospect of what will happen if they don't.

Arguably, the biggest drivers of our behaviours can be found in the myriad factors that influence people's desire to not cause discomfort on the one hand and to increase their own pleasure on the other. After all, people find safety in the comfort of doing what they have always done; they find relief in being a part of the bandwagon or not challenging and living outside of the perceived norms of where they live. They find reassurance in maintaining the status quo, of suppressing the dissonance between their values and actions, and alleviating any moral discomfort. Many people want to maintain their moral sense of self but do so comfortably and without the fear of rejection or ostracisation. So they decide to change their perception of their behaviour. In other words, they decide that they don't need to change at all. Their moral sense of self remains undisturbed because their actions are moral actions.

If only it were as simple as this.

TO BELIEVE, OR NOT TO BELIEVE

The question of how people reconcile what they wish to be true versus what is actually true is something that they must process internally. This question of believing or not believing is not just pertinent when viewed in relation to the evidence that exists, but it also relates to how people perceive themselves to be agents and authors of their own choices.

It's easy for both vegans and non-vegans to feel frustrated, confused and angry when we engage in conversations about our food choices. However, what we often neglect in these situations is the fact that the conversations we are having are not merely happening at surface level. There are significant layers of subtext – insinuations, biases, barriers, social and psychological influences, and the list goes on. We are like trees in a forest: we can easily see the parts above the ground, but that is only a portion of the whole – the foundations under the soil, a complex and interconnected system of roots that determine how the tree grows and allow it to exist in the first place, hold what is visible in place. It is these roots that hold up our values, beliefs and actions. It is what is out of sight that determines the very essence of how we act, think and believe.

The misinformation around the healthfulness of plant-based diets perfectly demonstrates the social influences and biases that influence people's behaviours. The political arguments relate to how people's identities shape their perceptions of veganism, and the arguments about taste and personal satisfaction show how difficult it is to challenge ingrained behaviours and how our agricultural system disconnects people from the suffering it involves.

So, as well as being aware of the drivers of peoples' behaviour, the question is: what can we do to ensure that we are being mindful of these drivers and, importantly, what tips and techniques can we use to ensure that we are arguing effectively? Which brings me to our next chapter.

BECOMING A MORE EFFECTIVE DEBATER: THE CORE SKILLS YOU NEED

Knowing what we want to say or what we want the other person to believe aren't always the hardest aspects of engaging in important conversations; instead, the most difficult part is knowing how to communicate what we want to say in a manner that will be effective and convincing.

As just discussed, vegans can sometimes suffer from an optics problem. This is partly a result of people unfairly maligning vegans so that they can shoot the messenger rather than deal with what we have to say, but is also partly because of how we sometimes communicate or present the arguments we are trying to make.

In many ways, the art of persuasion is not found in what we say but how we say it. If you're watching a debate and one person is fluent, confident, articulate and engaged, are they more convincing than their opponent if that person is withdrawn, doesn't make eye contact, stumbles over their words and seems unsure? The things is, in a situation like this, the charismatic speaker might seem more convincing, but that doesn't mean what they are saying is true or that they necessarily represent the correct position to take.

Perhaps there is another debate, and one of the speakers is

conscientious, listens, responds meaningfully with consideration and is likeable, whereas the other speaker is combative, angry, shouts over the other person and uses insults. Who are we more likely to listen to and agree with? Again, we might respond more positively to the person who is more likeable, but that doesn't mean that their position is the logically sound one.

In many ways, it is understandable that the person who is representing the stronger argument could get frustrated, angry and annoyed by a person who is flippantly and convincingly misrepresenting something. Imagine coming up against an effective speaker who is persuading people of something that is actually incorrect. Or imagine the disappointment that you might feel listening to someone convincing others to continue doing something that is causing suffering and harm to others. In such a situation, it would hardly be surprising if we became downtrodden by the fact that people were being persuaded that causing more harm was justified, especially if it was our job to convince them of the opposite.

This is why effective communication is not just about the content of what you are saying. And being aware of how body language and presentation can sway someone also makes us more effective critical engagers when we are listening to or debating other speakers who are persuasive and compelling.

So, how can we communicate more effectively?

ALL I KNOW IS THAT I KNOW NOTHING

First, learn your arguments. What is it that you believe and why do you believe it? How have you substantiated your belief and have you explored the arguments that those who disagree with you use?

In debate clubs, participants will often be challenged to take a position that they actually disagree with. This allows them to not only become better debaters, as it forces them to communicate ideas persuasively even when they are not convinced by them personally, but it also gets them to think more critically about the positions they do believe in.

How can we properly know that what we believe is true if we've never attempted to understand why people believe it is false? Exploring the arguments against something can often increase our conviction that we are right. This has been the case for me since I started debating and discussing veganism with people. In fact, even writing this book gave me the opportunity to really dig into what meat eaters try to argue, and by doing so has built up even more conviction within me about the merits of veganism. I hope this book is able to do the same for you.

The ancient Greek philosopher Socrates reportedly once said, 'All I know is that I know nothing.' This apparently paradoxical claim speaks to our lack of omniscience and inherent need for curious investigation. We are all ignorant of what it is that we do not know. I think such a view can also aid us when communicating with others, as it's important to engage in a conversation with the belief that there is something for us to learn from it too.

I'M ALL EARS

One of the simplest and most important ways that we can create more effective communication is by listening. Listening not only allows us to better understand what the person we are speaking to is saying; it also shows that we are sincere and respectful in our intentions.

While we don't have to respect the content of what someone

is saying, respecting their desire for a non-confrontational atmosphere and giving them the room to be able to express what they believe means that we can create a better environment for discussion, and by so doing allows us to challenge their argument in a more meaningful manner.

Shouting over someone or belittling them will not only make them listen to us less, but it also gives them more of a chance to avoid having to actually back up their arguments. If we remove the ability for someone to dodge a tricky question or to feel victimised, it creates a situation in which they actually have to engage with the question at hand.

Listening also allows us to ask better and more effective questions, and to directly address the points they are making. While this might seem obvious, all too often when we argue with someone we're thinking about what we want to say next, rather than listening so that how we respond actually counters what they are arguing.

It's also important to recognise that even if you don't understand what someone is saying or the point they are trying to make, asking them to repeat themselves or to further clarify what they are saying shows that you are listening to them, as an effective conversation is one where both parties are actively trying to understand the other person's perspective. Listening and asking for further clarification when needed means that we also reduce the risk of misrepresenting what the person is trying to say and refuting an argument that they are not actually making.

Another great way to show that you are listening is to repeat what someone is saying back to them, specifically using their choice of words. For example, drawing on an argument I referenced in the previous chapter, if someone says, 'I eat animal products because it keeps me connected to my cultural identity,' you could respond by saying, 'I understand that eating animal products keeps you connected to your cultural identity.' From that point you can then proceed to ask them questions such as: 'But is something moral just because it's cultural? And are there other ways that you can celebrate your culture that don't involve causing harm to others?' Which brings us to the importance of asking questions.

THE POWER OF QUESTIONS

Asking questions is also arguably one of the most effective tools that we have to draw out the contradictions in what people are saying. In fact, asking questions is so effective that Socrates was sentenced to death as a result of his skill in this regard, and because he was hated by certain sections of Athens for questioning their beliefs and therefore challenging their power. This form of questioning is now referred to as the Socratic method or Socratic questioning. It involves asking someone questions about their beliefs and then, rather than telling someone what you believe, asking them more questions based on their answers. Through this process of asking questions, we are able to probe for flaws in someone's logic and reveal inconsistencies, which in turn allows those involved in the conversation to reach a more robust and nuanced understanding of the issues being discussed. This form of questioning also allows us to draw out any assumptions that might have been left unexamined and challenge any biases or flaws in the person's argument, something that I was able to do when discussing veganism in Texas with someone called Kyle.

> **Kyle:** If I'm going to eat an animal, I want it to be treated respectfully up until the point it is killed and butchered.

Me: Sure, but at the same time, it's not respectful to harm someone unnecessarily. It wouldn't be respectful to beat animals, so why is it respectful then to cut their throats? How does that constitute respect?

Kyle: I don't have a good answer.

Me: Do you think that does constitute respect?

Kyle: I, yeah . . . I don't . . .

Me: What do you think? Just be honest – is it respectful to unnecessarily cut their throats?

Kyle: Respectful to a certain point.

Me: What point is that?

Kyle: I think they deserve a certain level of treatment up until the point that you're going to consume them.

Me: Why do they deserve to die? If they deserve to be treated well, don't they deserve to not be killed unnecessarily?

Kyle: [*Pause*]

Me: What do you think?

Kyle: No, I don't have something prepared for that. I really don't. I don't have a good answer or rebuttal.

In this situation, asking questions allowed me to draw out the inconsistencies in Kyle's arguments, and he ultimately recognised that he didn't have a good answer to what I was asking him. Also, by referencing a point he'd made earlier in the conversation about believing it to be wrong to hit an animal, I was then able to use his own previously stated belief to highlight the flaws in his logic. This is why listening and asking questions go hand in hand. By being attentive to the language and statements the person is making, we can ask even more pertinent and effective questions in return.

This exchange also highlights another important aspect of asking questions and leading someone to recognise the flaws in their position on their own. It is not often that we are able to sit down with someone and within 20 minutes get them to completely change their worldview. Conversations are more about planting new ideas in someone's mind or sowing doubt about their beliefs. While it would be great if we could talk to someone and have them change their mind on the spot, it's not realistic to expect that of them. Indeed, if we place too much expectation on ourselves, we will always feel disappointed about what we have achieved. Instead, our role as advocates is to leave someone with a better understanding of veganism and a more favourable view of it.

In essence, this is what it means to 'win' in this context. Our objective is to make rational and logical arguments for veganism and in doing so create more positive impressions about what being a vegan means. If we can achieve this and leave people with a more favourable perception of veganism, then we have achieved a 'win'.

In this discussion with Kyle, our conversation was amicable, polite and friendly, but at the same time I was able to challenge his beliefs to the point where he acknowledged he didn't know what to say. This acknowledgement was more important to him than it was to me. His inability to answer my questions revealed to him that his beliefs did not align with his actions.

Asking questions allows us to respond proactively to what the person is saying and take what they are arguing to its logical conclusion. In doing so, it also allows us to scrutinise circular logic and avoid repetitive statements. Without asking questions, a similar type of conversation to the one I had with Kyle could end up going along these lines:

Vegan: Eating meat is animal cruelty.

Non-vegan: I don't think that it is.

Vegan: How can you even say that? It is obviously animal cruelty. You're killing animals.

Non-vegan: I disagree. I don't think it is cruel because we use the animals for food.

Vegan: I don't eat animals and you don't have to either. It's wrong and cruel. Animals scream in pain and fight for their lives.

Non-vegan: I just don't agree with you that it's cruel, and I try and buy high-welfare animal products like free-range.

Vegan: Free-range doesn't mean anything. You're still paying for animals to die.

Non-vegan: Look, I respect your choice to be vegan, so you should respect my choice.

Vegan: I don't respect people who kill animals. Being vegan is not even hard.

And on we go . . .

Asking questions is also an effective way of removing an air of judgement or criticism from a conversation. Often people will feel challenged or get defensive when they are being told that they are wrong. After all, nobody likes to be told how they should feel or what they should think. We want to be able to make decisions for ourselves and come to our own conclusions. The Socratic method is a great way of avoiding telling people what they should think and instead invites them to analyse their beliefs and encourages them to think differently of their own volition. Questions that begin with 'why' are a great way of achieving this, as they are asking the person to substantiate their view and explain it.

MIND YOUR LANGUAGE

The words we use are also incredibly important when it comes to creating effective dialogue. Language has the ability to disgust, offend and divide. Yet it also has the ability to unify and inspire. So avoiding using aggressive or belittling words, and being conscious and deliberate with the language we use, is essential if we want to positively influence people.

One of the easiest ways to do this is to be mindful of how we use words like 'you' and 'your'. During a conversation with a meat eater called Cam, he declared that he 'definitely loves animals'. I could have said, 'Proclaiming that you love animals while still eating meat means you're a hypocrite,' but I instead said, 'Can you love them if you pay for them to be killed needlessly?' In response he smiled knowingly and said, 'See, that's a good question.'

In essence, I would have been saying the same thing. However, with the statement version ('Proclaiming that you love animals while still eating meat means you're a hypocrite') I would have been telling him something as a matter of fact, and the use of the word 'you' would have seemed accusatory. By framing my response as a question ('Can you love them if you pay for them to be killed needlessly?), the word 'you' was an invitation for him to answer based on how he felt. Alternatively, we can also use words like 'we' and 'our' instead to create a sense of togetherness and to highlight how these are collective behaviours and actions.

Being mindful of language works both ways, though, as we can highlight the other person's words and in turn use them ourselves. Questioning the language that the previously mentioned Kyle was using during our debate was helpful in reinforcing the point I was making:

> **Kyle:** I believe as people that we are above animals.
>
> **Me:** Are you above a horse?
>
> **Kyle:** Yes.
>
> **Me:** So why not kill the horse then?
>
> **Kyle:** Because I don't have to.
>
> **Me:** You don't have to kill the cows or chickens or pigs either.
>
> **Kyle:** To fulfil what I feel I need from my diet I do.
>
> **Me:** Need or want?
>
> **Kyle:** Yeah, that's fair.
>
> **Me:** Is it a need or a want?
>
> **Kyle:** I guess you're right.

Often people will use words to subtly reinforce their position. Kyle's use of the word 'need' was an attempt to provide a justification for why his views about horses compared to his views about cows, pigs and chickens were contradictory. However, by highlighting this and instead proposing the idea that he actually meant 'want', his contradiction was revealed to him.

THE IMPORTANCE OF VALIDATION

As well as being curious and asking questions, validating what people say where appropriate is a really strong way of signalling that you respect the person you are talking to and that you are listening and empathising with them. For example, if someone was to say that we need to eat animals for protein, you could respond with something like: 'I can understand why you would think that. After all, we're constantly told that meat is the main source of protein. But do you think we can get protein from plants?'

Validating is not necessarily the same as agreeing, although you can of course validate someone by acknowledging that you agree with what they're saying. However, validating someone even when you disagree with them is about making the person feel heard and an equal part of the conversation. It's about fostering an environment that reduces the potential for conflict and indicates good faith and sincerity.

People don't want to feel belittled or intellectually insecure, so validating allows you to handle when someone says something that is wrong without making them feel defensive. So, instead of saying, 'You're wrong,' you would say, 'I can understand why you believe that. However, have you considered . . .'

Demonstrating that you understand where someone is coming from and why they believe what they do can be a really simple yet powerful way of showing that you are listening to them and actively attempting to understand them. It ensures a person feels like they're being heard, from which point you can ask a question that gets to the root of why their argument is ultimately flawed.

IT'S NOT WHAT WE SAY BUT HOW WE SAY IT

So we've acknowledged that we need to be informed, and we've discussed the importance of listening and being mindful of the language we use. We've also started to ask more questions and begun to validate people's arguments when appropriate. But what else do we need to consider?

We've established that it's not necessarily what we say but how we say it that is important when it comes to making a convincing argument. This includes verbal elements, like the tone and loudness of our voices, but how we say something goes beyond the realm of just spoken language. We also need to consider our body language. We might be asking all the right questions and giving all the right responses, but if we're clenching our firsts, furrowing our brows, shaking our heads or pointing our fingers, we could well be undoing all the other work we're putting into the conversation.

When I first started advocating for veganism, whenever I started to get irritated or frustrated my index finger would emerge and start pointing at the person I was talking to – definitely not a good idea if you want people to feel comfortable and not defensive.

However, while visual cues can negatively impact a conversation, they can positively impact them as well. Facing towards the person we are speaking with and giving them a good amount of eye contact, particularly when they are talking, shows that we are committed to the conversation and that we are listening to them. And crossing our arms, fidgeting, tapping, putting our hands on our hips, checking our phones, gazing into space, sighing and other dismissive actions are all to be avoided if we want to create a situation in which the person we are speaking with feels comfortable and prepared to be honest and sincere with us.

Instead we can use our hands and arms to illustrate our points, and having relaxed shoulders and open palms can signal that we are calm and comfortable. Appearing at ease and having a calm demeanour are great ways of defusing tension, and having relaxed facial expressions, an open posture, and being mindful of not encroaching into the other person's personal space are all really helpful and effective tools too.

We can also use the pace of how we are talking to achieve the same thing. If the conversation feels like it is getting more tense and standoffish, taking longer pauses between sentences and words and opening our palms and arms shows a level of vulnerability and a desire to not let things spiral or get out of hand.

Smiling when it's appropriate can also create a more positive mood, as can nodding our heads in agreement. So if someone makes a valid or good point, something you agree with or understand, nodding or smiling can help the other person to feel confident and less like they are in an 'us versus them' scenario. Visually and verbally agreeing with someone shows that you are looking for common ground, and that the purpose of the conversation is to find out what points you both agree on and then dissect the ones you disagree about to find out why you don't align.

Seeking to find commonality is a really effective way of making an interaction more amicable and allows you to find out where your beliefs diverge. This can make our conversations more focused and relevant. It's not that uncommon to find ourselves in situations in which we are actually arguing about something we both agree about; yet, working to establish commonality can help to avoid such situations from arising.

However, being aware of body language also works the other way round – as well as being mindful of what signals our bodies might be giving off, it's also important to pick up on cues that are coming from the other person. Perhaps they are showing signs of being anxious – for example, they're avoiding eye contact or their voice is breaking a little or they're struggling to express themselves. To help them feel less anxious you could reduce the amount of eye contact you're giving them, slow down the rate you are talking, soften your voice so it's less intimidating and avoid interrupting them or talking over them, while at the same time encouraging them by emphasising that there's no rush and they can take as long as they need.

During a conversation a person might go from seeming to be relaxed to folding their arms and leaning away from you. This would be a signal that they have become uncomfortable or are feeling standoffish. In response you might try to validate their position more, or verbally acknowledge that it's a challenging conversation and that you appreciate them for engaging. You could also change the language you are using, as previously mentioned.

Shifts in a person's body language can also be good indicators of what the most effective questions we're asking actually are. During a conversation I had with a student in Arizona, he proclaimed while smiling, 'I love animals but I love the way they taste too.' He then paused, his smile dropped away and he furrowed his brow, as if confused by what he had just stated, before adding, 'So it's a bit contradictory.' His shift in body language indicated to me even before he acknowledged the contradiction that this was a point he was struggling to grapple with. As a consequence, I could tell that asking him about his love for animals would be an effective direction to take the conversation.

We can also mirror the body language of the person we are speaking to. This means subtly incorporating some of their body language and vocal qualities into our own. Mirroring is done to build up rapport and trust between individuals. So if the person we are speaking to adopts certain postures, adopting those same postures helps put them at ease. Likewise, if the person is speaking in a slow and calm manner, doing the same can be another effective way of mirroring.

LEAVING A GOOD IMPRESSION

But even with all that taken into consideration, sometimes conversations will spiral into more heated arguments. In such a scenario it's OK to end a conversation or to establish that it's no longer productive or helpful for either party. This is especially true with family or friends, where sometimes the best thing is to pick up the conversation again at a later date.

If you find yourself in such a situation, you could say, 'This is a difficult conversation to have, so I think it might be best to continue at another point once we've both had a chance to properly reflect on what's been said.'

One of the most important things to consider is that people remember how something made them feel. Often we have an experience or a conversation and we don't recall everything that was said, but we do remember how we felt as a consequence of it.

Think of a holiday you went on and loved – chances are you can't remember everything you did, all the food you ate, all the sights you saw, but you do remember that you had a great time. The same is true of conversations. You might go on a date or hang out with a friend and spend hours in their company. It's unlikely you'll remember everything you spoke about or every joke that they made, but you will remember that they made you laugh or that you enjoyed yourself.

This logic also applies to conversations about topics such as veganism. If we are having a lengthy discussion with someone, after the conversation has finished the chances we will remember every point that was made, or every question that was asked, or every response that was given, is low, especially as more time passes. However, we will remember how the conversation made us feel. Did we enjoy it, did we hate it, did we find it fun, did we feel pressured and judged? Perhaps we felt belittled and misrepresented, or perhaps we felt validated and came out of the conversation with a positive feeling about the person we were speaking to.

Ultimately, veganism is a topic of conversation or debate that can become heated, unproductive and argumentative. It encompasses so many important concepts, and calls into question what sort of people we believe we are, what sort of people we want to be, and what sort of world we want to live in. It is a conversation that can be passionate, emotional and is high-stakes. It does, after all, concern the lives and experiences of others, human and non-human alike, and the future of our planet and food. The stakes couldn't really be any higher.

As a result, it is sometimes easy for us to avoid the issue or not address it in the manner that it deserves. But it is too important to ignore, and engaging with the arguments and putting forward your points of view in a persuasive way is vital if we are to avoid widening the polarisation and deepening rift between those who are in favour of veganism, those who are not, and those who simply don't know yet.

So, as Archbishop Desmond Tutu once said, 'Don't raise your voice, improve your argument.'

V.E.G.A.N.

It's important to remember that not every conversation will go the way that we want it to. Even with our best intentions, a discussion can break down and become unproductive and hostile. On other occasions we might feel like we weren't able to communicate what we wanted to say properly, or we might think about it afterwards and wish we had said something different.

Effective communication is a practice, and like any practice, it's not about being perfect or faultless. We all make mistakes and that's perfectly normal. The trick is to try to always improve, to learn from our mistakes and to build up more confidence over time. The tips and techniques I have outlined in the first two chapters are tools that you can use to continue that process of improvement and confidence-building.

They are skills used in all industries where communication is vital, from politics to therapy to public speaking. And importantly, they are skills that will make you more effective at discussing veganism and give you the confidence to believe in yourself and your ability to positively inspire others.

To help you remember these skills, I have created an acronym that you can use. Handily, it is a word that should be easy to remember.

V – Validate. Make sure to be understanding of a person's beliefs and validate them in situations where it is appropriate.

E – Eloquent. Be deliberate and intentional with the words that you use, and be mindful of what you say.

G – Give attention. Make sure to listen to the person you are speaking with.

A – Ask questions. Scrutinise a person's beliefs and arguments by asking questions.

N – Non-verbal communication. Body language is a vital part of any effective communication, so be conscious of what your body is saying.

THE ARGUMENTS

The next chapters of the book focus on the arguments against veganism. As previously mentioned, these arguments have been grouped according to the different archetypes of people that I have come across. Because this book is a tool, you can jump between the archetypes and arguments in any manner that you wish. There are also cross-references throughout, so you can move through the book in the way that brings the most value to you.

My rebuttals focus on breaking down the arguments, providing context and discussing why they are not good enough justifications to not be vegan. So, as well as arming vegans with the knowledge they need to argue effectively with meat eaters, the arguments are put together in a way that means this book is useful for your non-vegan friends or family members too.

The list of arguments is extensive and exhaustive, ranging from the more obvious to the ones used by the animal farming industries themselves and the lobby groups working on their behalf. Throughout I investigate many of the studies that have been used as a stick with which to beat veganism and explain how these arguments have come about and why they are used. This means you will build up a deeper understanding of the arguments and become even more confident and assured in your rebuttals.

What follows is every argument against veganism.

THE MISTAKEN PHILOSOPHER

'MORALITY IS SUBJECTIVE AND/OR RELATIVE'

One of the main philosophical arguments levelled against vegans is that morality is subjective and/or relative. It's commonly used as a way of saying that without an objective morality, what we do to animals is down to the individual. You'd best bring your toboggan, because this is an awfully slippery slope.

If we are going to make the argument that causing harm to someone else is justified because morality is subjective and to each their own, then we have to apply that logic consistently, which of course we don't do. We don't even do that in the case of all non-human animals. Because to be consistent with that belief would mean that anything you do to an animal – torture, rape, mutilation, murder – is moral as long as the individual is OK with it. Most people don't take this position, especially as a world where morality was determined by the individual would also make it permissible to do terrible things to other humans.

Moral relativism is, broadly speaking, the belief that cultures and different peoples hold different beliefs, so nobody is objectively right or wrong. One of the most obvious issues around moral relativism is that by deciding that something is moral based on social and cultural norms, we take ourselves into dangerous territory (see 'The Social Conformer' for more on this). A good example of this came up during a conversation

I had with a student at UT Austin when I challenged him on his views of relativism.

> **Me:** Let's say there is a situation where ten people are harming one person and that one person can't stop them: are those ten people justified because they've decided that individually they want to and morally they don't have a problem with that?
>
> **Sahil:** According to those ten people, I would say that they are morally justified.
>
> **Me:** To harm them? To harm someone else?
>
> **Sahil:** Yes.

In the interests of not misrepresenting Sahil, he did also state that he didn't personally think it was moral. But this example perfectly illustrates the perils of moral relativism. This line of thinking can lead to things like slavery and apartheid becoming morally justified, as it essentially places what is moral in the hands of those who have the power. If the majority of people think something, and that something is therefore the cultural norm, that makes what the majority believe moral as a consequence. In the case of non-human animals, if you're a moral relativist you would have to think that the burning of animals alive was not immoral if it were taking place in a society where it was considered normal. I hope you still have that toboggan.

Some people believe that animals don't need to be involved in this conversation because they don't need to factor into our moral calculus. They would then need to explain why this is the case. They might make appeals to the intelligence of other animals (see 'The Egotist'), their religious teachings (see 'The Historian') or to things like social contracts (see below).

However, if we believe that animals should be included in our circle of moral consideration, then that means the way we treat them is an ethical concern. I strongly believe that the vast majority of us already do believe this. If we didn't, then raping, stabbing and mutilating animals wouldn't be a problem.

So if we establish that what we do to animals is an issue of ethics, that means that we then need an ethical justification to condone what we do to them. And the fundamental point of this book is to show that no such justification exists.

'IT'S BETTER FOR AN ANIMAL TO HAVE LIVED THAN NOT'

Let's say I breed a litter of puppies into existence. As soon as they are born I cut off their tails and extract their teeth. I then keep them in a concrete pen and when they're five to six months old I put them in a gas chamber. Is that moral? Because they wouldn't have been born if I hadn't wanted to do those things to them.

That entire sequence of events is what we do to pigs, so if the argument that it's better to have lived than not works in terms of justifying what we do to them, it should also work for other species as well. In fact, the logic of this argument should actually excuse doing anything to anyone as long as we brought that being whom we're harming into existence. Are parents allowed to abuse their children because their children wouldn't have existed without them and it's better to have lived than to have never lived?

Clearly we are not justified to cause harm and suffering, even if the animal wouldn't have lived without the existence of the industry that causes them that harm and suffering. But let's develop this point a little further.

While this argument obviously fails in a situation where an animal lives a miserable life, what if the animal lives a happy life and enjoys their existence? After all, if it hadn't been for them being bred to be killed for meat, they wouldn't have gotten to experience the happiness that they do.

This idea is often referred to as the net-positive theory. The American philosopher and author Sam Harris is a proponent of this excuse:

> I think there's an interesting ethical wrinkle with respect to the prospect of farming in such a way that the lives of farm animals are net positive. So that it's better to have existed as a cow than to have not existed at all. And if we had some sense that these are indeed happy cows, well then, it would be better to raise them and treat them well and then finally eat them.

We find ourselves at the top of yet another slippery slope if we think of what we might be able to do to others under the guise of a net-positive outcome. If a parent gives their child a great childhood filled with happiness and fun, does that make it permissible to hit them from time to time? The child could still be living a net-positive life, so does that make it OK? What about in the case of non-human animals? If someone has a dog, can they occasionally kick them as long as they take them for lots of walks and play fetch?

The problem with the net-positive argument is that it makes doing immoral things to others permissible as long as they are counterbalanced by positive things. However, hitting a child is still hitting a child. Kicking a dog is still kicking a dog. These things are immoral because of what they mean for the victims of those actions – being hit or kicked still causes suffering and harm regardless of the overall positivity in that victim's life. Furthermore, could we deny rights to someone, like a woman's right to vote, or a gay person's right to marriage, as long as overall the lives of women and gay people were positive and worth living?

From a practical perspective, it's important to recognise that the more space and welfare an animal receives, the less profitable and viable the farming becomes. Factory farming doesn't exist because our aim is to be as cruel as possible to animals. It exists because it's the only way to meet demand in a way that is even remotely financially or practically viable.

In the UK, the average sheep and cattle farmers would actually make an annual loss of more than £16,000 if it weren't for subsidies and government support.[1] And these are farms that carry out unethical practices and slaughter animals in industrial slaughterhouses. To create a life where suffering is minimised and overall wellbeing is increased as far as is possible for farmed animals would make the industry even less financially viable, leading to such a system being reserved only for those in higher socio-economic brackets.

The cruellest versions of animal farming, such as the farming of chickens and pigs, are actually among the most sustainable and financially viable. In other words, there is a direct trade-off. However, that trade-off does not exist when you eat plants, as not only is eating plants more ethical, it is also more sustainable. Not to mention that growing plants is more economical and efficient too.

But even if theoretically we could buy animal products from a farm where an animal lives a truly happy life, that doesn't make taking their life from them ethically justified, even if the slaughter is without suffering (see 'The Wishful Thinker').

You could also argue that if an animal lived a happy life, robbing them of their ability to continue living that happy life would be even more cruel. If we look at it from the opposite direction, a chicken raised for their flesh lives a horrendous life and, even though it is only six weeks long, it is filled with pain, suffering and distress. As such, death could even be considered to be a mercy for them, as it brings an end to their suffering. I'm not of course suggesting that we are justified in slaughtering chickens, as ethically we shouldn't be farming and causing them misery in the first place. But death couldn't be considered a mercy for those hypothetically happy cows. After all, each day that we don't kill them is another day in which they could continue to feel happy. How much happiness is enough happiness before we cut their throats?

Regardless of the positivity of their lives, if we are going to eat their flesh, we will at some point have to slaughter them, and in that moment our actions become unethical, in the same way that the moment we decide to hit a child who is otherwise living a happy life is when our actions become unethical. Plus, in order for the meat to not be chewy and unpleasant, we would still have to kill the cattle at a young age. Cows can live 20 to 25 years, but we kill them at 12 to 24 months because we prefer the taste of young flesh.

Let's say that you only have the ability to look after one cat at a time – is it best to look after that one cat and make sure they live a happy life for as long as possible, or should you kill your cat every six months so that in the space of 15 years you could have 30 cats living a net-positive life rather than just one?

When it comes down to it, what we should really be saying is that it's better for an animal who is alive to be given the happiest life possible, and that means one in which they're not exploited and slaughtered.

'IF WE ALL WENT VEGAN, THE ANIMALS WE FARM WOULD GO EXTINCT'

The first very simple way to counter this argument is to establish that animal farming is the leading driver of species extinction. When we analyse agriculture, including the farming of fish, 24,000 of the 28,000 species that are currently facing extinction are doing so because agriculture poses a threat to them. And in marine ecosystems, the largest driver of biodiversity loss is the fishing industry.[2] Basically, eating animal products because you are worried about the extinction of animals is akin to driving a pick-up truck because you are worried about the greenhouse gas emissions of cars.

The next thing to note is that the animals we farm have been selectively bred and are domesticated agricultural animals. Let's say that dog fighters used a breed that they had selectively bred for dog fighting and that outside of the dog-fighting world this type of dog didn't exist. Should we continue dog fighting because without it this species of selectively bred dog would become extinct?

Wild chickens, wild cattle, wild sheep, wild turkeys, wild ducks, etc. still exist in the world. We would not be eliminating all of these animal species – we'd simply no longer breed animals that wouldn't have existed without agriculture. Sure there would be fewer cows, pigs and chickens overall, but there would never have been this many of them if it weren't for farming. Of all the non-human mammals on the planet, 96 per cent of the global biomass is farmed mammals compared to only 4 per cent being wild mammals. In the case of birds, 71 per cent of global bird biomass is farmed birds, with only 29 per cent being wild birds.[3] By reducing the number of farmed animals that exist, we would be providing an opportunity for more animal species to exist overall.

Plus, if we really wanted to keep these selectively bred animals around, we could still do so – they could live in sanctuaries instead of factory farms and not be slaughtered. That is, of course, if the real reason we're concerned is actually because of the extinction of these animals. In my experience, the reason people use this argument is not because they're genuinely worried about the extinction of beef cattle – it's because they're worried that no beef cattle means the extinction of steak for them.

'AS LONG AS WE USE EVERY PART OF THE ANIMAL AND DON'T LET THEM GO TO WASTE, KILLING THEM IS ETHICAL'

There is a pervasive myth that we don't utilise the whole animal, but in truth we do. We boil the bones to make gelatine. We use the blood for additives, pharmaceuticals, fertilisers and animal feed, to name but a few applications. We take the parts of the animals that are not going to be sold for human consumption and render them down to be used as pet food.

The slaughtering of animals is many things but wasteful is not one of them. Does this make what we do to animals ethical, though? To draw a human comparison, the serial killer Ed Gein used the skin of the people he murdered to upholster chairs and made bowls out of their skulls. Does that make what he did any less immoral?

The reason what we do to animals is unethical is because of what it means for the animals. It's important that we view issues of morality from the perspective of the individual who is being harmed and exploited. Let's take fast fashion as an example. Is forced labour in fast fashion moral as long as we actually wear the clothes? No, because a human has still been exploited to produce that item and their suffering is exactly the same. Is it wasteful if we buy but don't wear the clothing? Of course it is. But just because it's more wasteful to buy the clothing and not wear it it doesn't make buying and wearing it ethical as a consequence.

The same logic is true with meat, dairy and eggs. While it is less wasteful if we use the whole animal, that doesn't mean it is ethical to do what we do to them in the first place. Once an animal is dead, they're dead. Using their whole body doesn't make them any less dead, and it doesn't mean that they suffered less while they were alive.

'ANIMALS CAN'T ENGAGE IN SOCIAL CONTRACTS'

The social contract theory states that rights are given to agents who agree to act and behave in a mutually beneficial way. For example, we might create a mutually beneficial system of rights and responsibilities in which I agree not to murder you, so you in return agree not to murder me. The logic then goes that because non-humans are unable to enter into social contracts, they are not therefore deserving of the protections that come as a consequence of them.

During a segment named 'Animal rights and wrongs' on a show called *The Agenda*, the author Jordan Peterson used this argument when referring to the aims of animal rights activists: 'They [animal rights proponents] don't care that animals cannot reciprocate their so-called rights even though such reciprocity constitutes a cornerstone of the social contract.'

But is it wise to assign rights based on the ability of those receiving them to reciprocate? A human with severe cognitive difficulties might be unable to enter into a social contract, as they are not conceptually able to reciprocate. Would that mean that we could refuse them their rights as a human? If an elderly person developed severe Alzheimer's and was no longer able to partake in social contracts, would they now forgo their rights? If we still believe that rights should be granted to these individuals, then the idea of social contracts and reciprocal rights falls apart. In reality, the way we structure our society is not based on Peterson's logic, because the rights we have to protect our wellbeing and safety exist even if we don't reciprocate or engage in social contracts ourselves.

On what basis are rights being assigned then? If it's wrong to cause suffering to someone if they are unable to agree to a social contract, what does that tell us about inflicting pain on others? If it's the capacity to suffer itself that makes causing pain wrong, does that mean that non-human animals who suffer should be safeguarded, regardless of whether they can engage in social contracts?

The theory of social contracts is also political in the sense that, for the most part, social contracts exist between the groups that make up a society and the government. But this opens up the door to discrimination of minorities. If Black people are not viewed as 'whole' humans and are thus not treated as moral agents in the same way as white people, are they protected by social contracts? History shows us not. The reason slavery in the USA ended was not because of a mutually beneficial and agreed-upon social contract. It took a war. And even then, racism didn't end. So social contracts fail in a human-only context anyway.

Sometimes the granting of rights is not mutually beneficial either. The right to not be a slave is not mutually beneficial to someone who is a slave owner. So, if the social contract theory posits that rights are granted based on agents agreeing to act in a mutually beneficial way, how do civil rights factor into this? Is the argument that before civil rights both white people and Black people were negatively impacted by the denial of rights to Black people, hence why these mutually beneficial rights were brought in? Clearly not. If slavery had not been beneficial to white people, it would have never existed in the first place.

Who decides what social contracts should exist? Obviously this falls on those who have the power – it is a form of 'might makes right'. In other words, because someone has the strength to create and enforce social contracts, the social contracts they create are the ones that are enforced. This significantly hampers minorities and those who lack social power.

It is often falsely assumed that the pursuit of granting animal rights is an attempt to give non-human animals the same rights as humans. But the idea that non-human animals and humans should have the same rights is of course non-sensical, and this is simply a straw man argument. The right to vote is pointless for a pig or a chicken. Animal rights is about pursuing rights that are relevant for animals. For example, the right to not be slaughtered is a very applicable and relevant right for an animal. So instead of being concerned about whether non-humans can reciprocate rights, we should instead be concerned about the ethical implications that come as a consequence of us denying them those rights.

THE EGOTIST

'ANIMAL PRODUCTS TASTE GOOD'

Does our pleasure morally justify what we do to others? Because essentially that's the argument that is being made here – that what we do to animals is acceptable because their flesh and secretions are tasty. Obviously this logic could lead to all sorts of horrific actions being deemed morally permissible, as long as the person enjoyed what they were doing.

We also don't use this argument consistently when dealing with issues concerning non-human animals. And yet, this argument is probably the one that is used the most to justify not being vegan, despite the fact that most of the people who use the excuse would be horrified if its logic was extended to justify many of the other terrible things that are done to animals in the name of enjoyment.

It is of course true that animal products can taste nice. I used to love the taste of them too. But part of going vegan is recognising that our sensory pleasure is not more important than the lives and experiences of the animals who are killed to satisfy our tastebuds. Plus, we're talking about a fleeting moment of time. A meal lasts about 15 minutes. We take animals' whole lives from them, forcing them into an environment that robs them of their autonomy and causes them suffering, before we then cut their throats or force them into gas chambers. And all of this for a matter of minutes. So much anguish, destruction and suffering for a transient moment of tastebud stimulation. A meal that we then forget about. Is this not utterly absurd?

And going vegan doesn't mean you stop enjoying food. Plant-based food can taste delicious as well.

'HUMANS ARE SUPERIOR/MORE INTELLIGENT SO WE CAN DO WHAT WE WANT TO ANIMALS'

In order for this argument to hold, there would need to be something intrinsic about humans that makes us not only superior but so much more so that our desires and wants are viewed as more important than the lives and experiences of other animals. So, what is it about humans that makes us so superior?

Most of the time, the answer to this question is that humans are more intelligent than other animals. But what about human babies, who are less intelligent than many animals? Can we kill them? Of course not. Someone might counter that a baby will become more intelligent. But what if they didn't? What if they had a serious cognitive condition that limited their intellectual capacity and meant they couldn't solve puzzles or make conscious decisions in the way that many animals can? Would we then be justified to treat them as we do animals because they lack the intelligence that other humans have?

If the answer to those questions is no, then intelligence isn't the source of human superiority or a justification for our mistreatment of animals. In fact, babies, children and elderly people with a cognitive condition like dementia are viewed as warranting extra care, support and consideration. The act of harming someone more vulnerable than us in a human context is viewed as being among the most abhorrent and evil things one can do.

Plus, the logical conclusion to this argument would be veganism anyway, as plants are the least intelligent form of edible life and would therefore be the most moral option we can choose.

I often hear people say that 'animals can't comprehend death' or that they 'don't understand the concept of time'. While there are good indicators that animals do have an understanding of death, such as showing signs of mourning, grief and loss, and they also exhibit fear in slaughterhouses, even if they don't comprehend death and the loss of time and life, could this not potentially make what we do even worse? Let's say you're a pig on a farm. You've been mutilated, forcibly impregnated, trapped in a cage you can't turn around in, and this is all you've ever known. How do we know that that pig doesn't think that this is all they will ever know?

For that pig, without the awareness that what is happening to them could ever end, and without comprehending death, it may well seem like this is what will happen for eternity. They don't know about a life outside of that farm, they can't conceptualise what being free would look like, because for all they know there is no such thing.

Is this not how we describe hell? A place of eternal subjugation, fear and suffering? And if so, have we not literally created a hell on earth for these animals?

Because humans are more intelligent and have an understanding of how experiences like pain and suffering will end, either because of medicine or death, we then know that it will stop. Plus our intelligence allows us to distract ourselves in ways that maybe animals can't. Sometimes I hear that animals just live in the moment. Well, only being able to live in the moment makes suffering so much worse. Using this logic, they can't distract themselves, they can't think or plan for a future without the experience they're enduring. If anything, many of the justifications we use to demean the experience of animals could actually make what we do to them even worse.

We might use traits like intelligence as a determining factor in a situation where we have to choose either/or – such as choosing to save either a pig or a grasshopper – but that doesn't mean going around killing all the apparently less intelligent beings we can find is ethical as a result. Similarly, if you had to choose to save a human child or a 90-year-old human, there would be a strong case for saving the child. That doesn't mean, though, that we can then go around killing everyone who is 90 years or older.

The truth is, we're not faced with the choices just outlined. But we do have the choice between animals or plants. And based on what we know about animal intelligence and sentience, and considering that we have to eat something, eating plants instead of animals reduces the harm that our diets cause and is the ethically preferable choice irrespective of how highly we view our own species.

The 'humans are superior' argument again ultimately boils down to 'might makes right'. But just because we can physically do something doesn't mean we should. In fact, having the power to subjugate someone else only increases our responsibility to act with compassion and care towards those more vulnerable than we are.

'I KNOW ANIMALS SUFFER BUT I JUST DON'T CARE'

I think to most people this point of view is concerning, to say the least. Thankfully, it is not a position that many people truly hold – in my experience, even people who eat animal products like to think that they're not causing suffering, although they sadly are.

I always wonder if a person would accept this logic if someone else were using it to justify harming them. Because this logic can be used to justify causing suffering to literally anyone. People who think this way do so from the comfort of having a safety blanket of social norms and laws that reduce the risk of people doing awful things to them. It's easy to take this position when your wellbeing and life aren't being threatened.

Stating that you simply don't care about the suffering of others is a cop-out. It's an easy way of putting up a barrier and removing accountability. It's basically saying, 'Don't even bother trying to convince me of the logic and virtues of your arguments because you won't get anywhere.'

The reason some people might say they don't care about animal suffering, without necessarily truly believing it, is because as soon we recognise that it is something we should be concerned about, it creates an obvious conflict between what we think and what we do.

'ANIMALS ARE NOT A SOMEONE'

The language we use to describe others is incredibly important. Language can denigrate, demean and belittle, or it can raise up, empower and equalise. The way that we refer to animals is therefore instrumental in shaping how we view and ultimately treat them.

For many of us, the idea of referring to animals as a someone rather than an *it* might feel odd. Idean, a student at UC Berkeley, had that initial reaction.

Idean: Why would you limit yourself to a certain amount of flavours when there's an exponential amount of flavours to try?

Me: Because someone has to suffer and have their life taken from them for some of the flavours that we're talking about.

Idean: It's not a someone.

Me: Are animals not someones?

Idean: No.

Me: Then what are they?

Idean: They're animals.

Me: Well, if an animal isn't a someone, then what are they? A something?

The word 'someone' is used as a term to define an individual. Animals are not objects, and they're not inanimate possessions – they are living, breathing, conscious, sentient individuals. They have subjective experiences, and feel pleasure and happiness, fear and pain. What we do to them has a tangible effect on their experiences and on their lives.

It's important that we view animals as individuals and recognise that when we talk about killing trillions of animals, we're actually talking about the experiences of trillions of individuals, each being forced to endure the reality that we force upon them. Each chicken in the barn of 40,000 chickens is feeling the burning of the ammonia. Each mother pig confined in a farrowing or gestation crate is experiencing the panic and anxiety of being confined, unable to move, surrounded by metal and concrete.

The thing is, we can't empathise with abstract concepts and amorphous masses made up of huge numbers, but we can empathise with the experience of an individual. Perhaps this is why we use demeaning language when we refer to animals? Hiding behind the facade of thinking of them as automatons, things, objects, makes it easier for us to ignore their suffering.

If we deny them the right to be viewed as a someone, we remove the need to identify with them. We remove the need to think about *them*, because we don't view them as a *them*, we view them as an *it*. Our language makes them comparable to the objects we own and not comparable to us.

Yet they are comparable to us. They, like us, are sentient beings. They, like us, have families and friends. They, like us, wish to avoid pain, suffering and terror. They, like us, are someones, not somethings, and they deserve to be treated as such.

THE WISHFUL THINKER

'ANIMALS DON'T DIE FOR MILK OR EGGS'

This is a common misconception and also an understandable one. After all, these products are secretions and can only be produced by an animal who is alive, so the idea that animals would be killed in their production seems contradictory.

But even though it could be argued that animals aren't killed specifically to acquire these products, they are killed as a consequence of them. Take dairy cows, for example. Due to the burden and strain placed on their bodies from being repeatedly impregnated and selectively bred to produce up to ten times more milk than they would naturally, they become infertile, develop infections like mastitis or simply start producing less milk. This leads to the farmer being faced with two options: either keep the cow alive, paying for food and bedding while making less or even no money from the cow, or have the cow slaughtered, get paid for their flesh and then replace the cow with another who is fertile and will produce the most milk possible.

The farmer has no choice but to go with the second option if they want their business to be viable, so every dairy cow is ultimately slaughtered, and even though cows can live between 20 and 25 years, dairy cows are slaughtered as young as four years old if the farmer stops viewing them as being profitable enough to keep alive. And this doesn't even take into account the male calves who, useless to the dairy industry because they don't produce milk, are either slaughtered when they're days old, sold into the beef industry to be slaughtered when they're about a year old, kept alive for 16 weeks or so and then killed for veal, or live-exported to be killed in a different country. In other words, if you're an animal bred into the dairy industry, it's not a question of whether you'll be killed, it's a question of when.

When it comes to egg-laying hens, because they've been selectively bred to produce almost an egg a day, compared to the 12–20 eggs a wild jungle fowl lays in a year, their production declines after about 72 weeks due to the strain being placed on their bodies. So what do farmers do? They sell them so they can make money from their flesh and then restock their farms with young hens instead. The male chicks, who are useless to the egg industry because they don't lay eggs and useless to the meat industry because they haven't been selectively bred to grow to slaughter weight in six weeks, are killed as soon as they are born by being gassed to death or macerated alive.

This is the fundamental contradiction of vegetarianism: it's a lifestyle that exists because people don't want to pay for animals to be slaughtered, yet it continues to support the very thing it is supposed to be against – animals being slaughtered. In fact, dairy cows arguably suffer more than beef cattle, as the latter are more likely to graze in fields and don't have to endure the same emotional distress that dairy cows do or the physical burden of being selectively bred to produce so much milk.

While I think it's great that people make changes to try and reduce their harm to animals, vegetarianism has more in common with eating meat than it does with veganism. And upsettingly, for all of its good intentions, replacing beef and lamb with dairy and eggs could actually increase the number of animals suffering and being slaughtered.

'THERE ARE BAD APPLES BUT ANIMAL FARMING ISN'T LIKE WHAT WE SEE IN EXPOSÉS'

This is a comforting belief to have. After all, who wouldn't want to believe that animal cruelty is a rarity and not systemic? I think we have all believed this at one point or another, or at least attempted to convince ourselves that it is true. But is it?

Exposés and investigations shatter the facade of ethical animal farming, but all too often we just express outrage at the farm featured instead of at the fundamental issue at the heart of what is being depicted. We fail to see the wood for the trees. While exposés shine a light on what happens to animals on a particular farm or by a particular company, more often than not the acts shown are not unique to the places being exposed. Even when we just look at the legal, standard practices we can see that animal farming is not a case of a few bad apples acting cruelly – mistreatment and abuse is systemic.

Male animals are coerced into ejaculating. Female animals are penetrated and forcibly impregnated. Newborn animals are mutilated, castrated and then taken away from their mothers. Animals are kept in cages, concrete pens and ammonia-riddled barns. They're selectively bred to make them grow larger, to produce more milk and eggs, or to have more babies, all of which negatively impacts their bodies and wellbeing. They're shot, have their necks snapped, and are slammed against floors and walls for being too weak or for not growing fast enough. They're crammed into small boxes or forced into trucks that take them to facilities that exist with the sole purpose of killing them. They're herded onto kill floors, into metal boxes or into gondolas in gas chambers.

They have bolts fired into their skulls, electricity forced into their brains or carbon dioxide pumped into their lungs.

And all so that we can then pull a blade across their throats, drain them of the blood in their bodies and cut them open, pull out all of their organs, cut off their heads, slice up the parts we want to wrap in plastic before then rendering down everything else that's left.

This isn't just factory farming. Remove the cages and concrete pens, and animals raised outdoors still face these horrors as well. They still have their tails docked and rubber bands wrapped around their testicles. They're still ear-tagged, still selectively bred, still impregnated. Does an outdoor pig farmer spend money and time on a sick and ill pig who won't make them any money, or do they kill them on the farm just like the pig farmers who use farrowing crates? Factory farming might be worse, but just because some animals get to feel the sunshine doesn't make everything else that we do to them moral as a consequence.

And then, of course, there's still the slaughter. Every animal farmed meets the same fate. Is a pig in a gas chamber suffering less because they weren't forced to give birth in a cage? Does a cow about to have their throat cut feel less fear because they got to graze on grass? Do the lifeless eyes still attached to their severed heads depict any less anguish from the final moments of these animals?

The fact that we want to ease our conscience by clinging to the false belief that animal welfare is high, and examples to the contrary are the exception rather than the rule, speaks to something important. We do care about animals. However, we end up in a conflicted state where we care about them but also don't want to change how we live or recognise that we are engaging in something that contradicts our values.

It's no wonder then that we fall into the trap of believing the humane-washing and idealised untruths that are sold to us. Because the real truth is that cruelty in animal farming is not a case of a few bad apples acting outside of the norm – the whole tree is rotten from the roots to the fruit itself. We can't just pick off the dead leaves and expect the problem to fix itself. There is a fundamental disease in the tree, and the only way to stop it is to plant a new tree instead.

'KILLING ANIMALS HUMANELY MAKES EATING MEAT ETHICAL'

In the context of slaughtering animals, the word 'humane' is commonly used to mean 'without suffering'. However, while killing without causing suffering is preferable, it does not make the act of killing moral in and of itself.

Sometimes euthanasia is used as an example to show that it is possible to kill an animal in a humane way, yet euthanasia is done in the best interests of the animal. After all, an animal is euthanised to relieve their suffering and because their life has reached a point where it is no longer preferable to live it.

But a farmer doesn't send an animal to slaughter because it is in the best interests of the animal – they do it because it is in their own best interests. We don't eat the flesh of an animal because we think that's the most compassionate way to treat an animal, hence why we let our pets live as long as they possibly can and don't cut their throats when they're six months old. Imagine what people would say if someone did kill a dog who didn't want to die and who wasn't suffering. Except, we don't need to imagine.

In 2021, YouTubers Nikki and Dan Phillippi decided to euthanise their healthy nine-year-old dog Bowser, not because he was suffering but instead because he had nipped the

Phillippis' son when he had removed Bowser's food bowl while he was eating.

Bowser's story epitomises the happy animal who lived a good life, but this still didn't dampen the reaction of people online who called the YouTubers murderers. Not to mention that Bowser's death was significantly more humane than any death that has ever occurred in a slaughterhouse. In fact, Bowser was killed in the most humane way possible. However, Bowser's death couldn't be considered humane regardless of the fact that he was killed in the best way possible because he didn't need or want to die. The same is true of farmed animals. Their lives are taken from them needlessly and against their will, and to make matters worse, they are forced into gas chambers and macerators or hung upside down so a blade can be dragged across their throat. Imagine the reaction if Bowser had been killed in a slaughterhouse. But if slaughterhouses were humane, wouldn't we want our pets to be euthanised in them?

A truly humane death is one that is done in the best interests of the individual who is being killed. So to kill someone who does not want to die or without acting in their best interests is an immoral thing to do, even if the death did not involve suffering.

And this would be true even if animal slaughter was without suffering, which it isn't. Even the slaughter that we call humane involves animals who are not stunned properly, have to be stunned multiple times or regain consciousness as they're being killed. Even if the slaughtering goes exactly to plan, there is still the fear and anxiety that the animals experience, with the sounds, smells and sights around them contributing to their emotional distress.

To understand why the issue of slaughter transcends just suffering at the point of death, we need only place ourselves in the position of the animal. Something we should always do when attempting to rationalise how we treat them.

Plus, let's say we opened a thesaurus and looked for synonyms for the word 'humane', what would we find? 'Compassionate', 'benevolent,' 'kind'. Is taking the life of someone who doesn't want to die compassionate, benevolent or kind? Similarly, if we looked for synonyms for the word 'slaughter', we'd see 'bloodbath', 'massacre' and 'murder'. How do 'compassionate' bloodbath, 'benevolent' massacre and 'kind' murder sound? The reality is that 'humane slaughter' is an oxymoron.

'THE ANIMALS I BUY HAVE ALREADY BEEN KILLED'

With the exception of animals like lobsters and crabs, which we can sadistically choose to be killed and served to us in some restaurants, it is of course true that in Western cultures, generally speaking, the animals we buy to eat have already been killed. But why have they been killed, and what does buying that meat mean for other animals in the future?

Even though the choosing of a specific lobster might instinctively feel like someone is more directly involved in the slaughter of that animal, it is really only our own sensibilities that make it any different from buying an animal that has already been killed. Ultimately we're still paying for the same thing – an animal to be slaughtered for us to consume.

It's a case of supply and demand. When we buy a product, we create demand for that product to be resupplied. So yes, the animal we are eating is already dead – it's just that another animal is now going to die too because of the choices that we have made.

Going off on a bit of a tangent, people often use a similar

line of thinking when they claim that because they're not kill-ing an animal themselves, it means they are not morally responsible for the killing. Which is good news for anyone who wants to have someone killed. Just pay for a hitman to do it on your behalf and you're not morally responsible. Because that's what this argument is fundamentally saying. A slaughterhouse worker is essentially a hitman whom we've hired to kill an animal on our behalf. The reason a worker slaughters an animal is because we have paid for them to do so through the purchases we have made. The blood of the animal might literally be on their hands, but it is our hands that are actually the most blood-soaked.

'IF I GO VEGAN, IT'S NOT GOING TO STOP ANIMAL FARMING'

What if everyone used this argument? It's easy to make an appeal to futility (which is the argument that because the proposed idea isn't perfect it's therefore not worth attempt-ing) or to remove any accountability by saying that you alone won't solve all of the world's issues, but that doesn't mean we are justified in continuing to be a part of the problem.

One person turning down an item made with single-use plastic is not going to solve the issue of plastic pollution, but that isn't a reason in and of itself for them to buy it. And what is a movement but a group of individuals all doing something as individuals to create change? The reason we have the system of farming we have now is because individu-als support it through their daily choices, but if those same individuals stopped supporting it, then it would cease to exist.

Making positive choices in our lives is not about believing

ourselves to be the sole person who can change something in its entirety, but instead recognising that we have a responsibility to play our part in helping to bring about that change.

'FISH DON'T FEEL PAIN'

It would be comforting to think that fish don't feel pain. After all, if we were able to convince ourselves that they didn't, it would certainly make it easier to reconcile the fact that we kill so many of them.

It's estimated that as many as 167 billion farmed fish are killed each year,[1] and the number of wild fish dwarfs even that number, with somewhere between 800 billion and 2.3 trillion wild fish being pulled out of our oceans annually.[2] Scarily, that number only reflects the legally recorded catch and doesn't include illegal, unregulated or unreported fishing, which some people estimate could constitute about 20 per cent of the global fish catch.[3]

With so many fish killed each year, what happens to them should be of real concern. Farmed fish are confined in overcrowded pens that make them susceptible to diseases and parasites that can cause spinal deformities, blindness, fin damage and death. Data from the fish farming industry and the UK government showed that 15 million salmon died on Scottish farms alone in 2022, and 1 in 4 die before they reach slaughter age.[4] The ones that do reach slaughter age are killed by being gassed to death, being left to suffocate in air or on ice, or by being bled out.

Wild fish are normally caught using either longlines, bottom trawling or gill nets. Longlines and gill nets can involve the animals being entangled or impaled for hours, leading to them suffering from their wounds for long periods and even suffocating. Once pulled out of the ocean, the fish can die from being crushed under the weight of the rest of the catch or from their organs rupturing from the change in pressure. Those that are still alive when they are landed will die from suffocation or by being eviscerated, meaning they are gutted alive.

It's no wonder, therefore, that we like to tell ourselves they don't suffer or feel pain. After all, paying for animals who can feel pain to be gutted alive would certainly be a challenging action to justify.

So, can fish feel pain or not?

To answer that question, a 2011 study into the responses of zebra fish is a good place for us to start. A tank was set up in which fish were placed in a column in the middle that then opened up to two separate tanks on each side. In one of the side tanks was a range of gravel, plants and fun things for the fish. The other was a barren tank.

To begin with, all of the fish went to the side with the enriching items. Half the fish were then injected with a type of acid that would cause them pain, if they could in fact feel it, and the other half with a saline solution, which was harmless. Again, all of the fish went to the side with the enriching items. Finally, half of the fish were injected with the acid and the other half with the saline solution. However, this time the barren tank contained some diluted painkiller. The fish injected with acid went to the side with the painkiller, meaning they sacrificed the enriched tank for pain relief.[5] If they were unable to feel pain, why would these fish forgo their preference?

There are some dissenting voices who think that fish don't contain complex enough cortical regions in their brains to feel pain and that their behavioural responses are not therefore an indication of a subjective experience of suffering. However, the general consensus is that fish do indeed feel pain. The desire for pain relief and the intelligence that fish display when they learn to avoid experiences that are designed to cause them harm are strong indicators that they do in fact suffer, a position supported by a review of 98 studies on fish pain.[6]

There's a fundamental evolutionary reason why fish would feel pain, which is that pain and the fear of it are incredibly important in ensuring that we protect ourselves from harm and potential death. It is pivotal in ensuring we learn from our mistakes, are wary of the dangers present in our environments, and fear predators and those who could do us harm. For fish, the evolutionary reason they feel pain is the same as our own and the same reason why pain exists for other animals as well.

It would of course be easier if fish were indeed just animated sea legumes. However, that's not the case. They too are sentient, feeling beings, and we can't hide from the fact that hundreds of billions of them suffer every year because of what we do to them.

'WHY DO VEGANS WANT TO EAT FOODS THAT LOOK AND TASTE LIKE ANIMAL PRODUCTS?'

I've never met someone who went vegan because they all of a sudden stopped enjoying the taste of animal products. I am certainly no exception, as I used to really enjoy eating animal products as well, especially fried chicken, steak and halloumi. The truth is, animal products are tasty.

However, as discussed previously, taste does not provide a moral justification for what we do to animals, so vegans go vegan because they recognise that life is more important than taste. That being said, taste is one of the biggest arguments used against veganism and one of the primary reasons why people eat animal products. So, by creating tasty plant-based alternatives, it means that people can go vegan and not sacrifice the flavours, textures and satisfactions that they have historically gotten from animal products.

It's basically a win-win, whereby we can still get the things we enjoyed from animal products but also reduce all the negative, unsustainable and unethical aspects of our food choices at the same time. Plus, with vegan alternatives improving all the time, the future of food innovation and animal-free food technology is incredibly exciting.

'MEN SHOULDN'T EAT SOYA AS IT REDUCES TESTOSTERONE AND CAN CAUSE MAN-BOOBS'

When it comes to influencing people's food choices, there are a number of things that are particularly effective. You might think that pandemics, antibiotic resistance, chronic disease, environmental collapse and animal suffering would be at the top of the list, but for some people there is something even worse and more scary than all of those things: man-boobs. In fact, legend has it that if a man stands in front of a mirror and says 'tofu' three times, his testosterone levels plummet and he suddenly develops man-boobs.

All jokes aside, the reason why soya is feared to cause the feminisation of men is because it contains phytoestrogen, an oestrogen-like compound found in plants. However, phytoestrogens work differently to animal oestrogen and have a much weaker effect. They are also selective, which means they have pro-oestrogenic effects in some tissues and anti-oestrogenic effects in others. This is why the consumption of phytoestrogens can actually help improve things like bone strength and density, while at the same time lowering the risk of the dangers associated with animal oestrogen consumption. This is especially important for menopausal women. A large-scale analysis revealed that menopausal women who consumed isoflavones, which are a type of phytoestrogen, had significantly increased bone mineral density when compared to women who did not.[7] Soya intake has also been associated with a reduced risk of breast cancer.[8]

But what about men? A 2021 analysis of 41 studies looking at the impact of soya protein and soya isoflavone intake on male hormones showed that 'regardless of dose and study duration' neither had effects on hormone levels.[9]

So where does the fear of man-boobs come from? Well, a lot of it stems from the case of a 60-year-old man who developed gynaecomastia, the clinical term for man-boobs. It was revealed that he was drinking 12 servings of soya milk a day and when he stopped doing so his 'breast tenderness resolved'.[10] That's it. One man who was consuming huge amounts of soya every day. Plus, an underlying sensitivity to isoflavones wasn't ruled out either, meaning that even if the excessive soya consumption triggered his gynaecomastia, it didn't prove that excessive soya consumption would cause gynaecomastia more generally in others. Not exactly the smoking gun pro-meat enthusiasts are looking for.

Another review of nine studies found that even at levels considerably higher than are consumed by Asian males, the biggest consumers of soya products, isoflavones did 'not exert feminizing effects on men' or result in breast growth.[11] And a study carried out on children who were being raised on soya protein formulas (SPFs) showed no early puberty, no changes in their bones, no gynaecomastia and no indicators of hormonal imbalances. They concluded 'that long-term feeding with SPFs in early life does not seem to produce oestrogen-like hormonal effects'.[12]

As well as a misconception about soya causing man-boobs,

some people fear that it will also reduce fertility. This anxiety mainly stems from a study of 99 men that found soya intake was associated with lower sperm concentration.[13] However, when adjusted for variables such as ejaculate volume, which was higher in those with the most soya intake, and when total sperm count was analysed rather than sperm concentration, there was no statistically significant decrease detected. Plus, the same research group did a follow-up study on almost twice as many men undergoing in vitro fertilisation with their partners and found that intakes of soya food were unrelated to fertilisation rates.[14]

The evidence shows that soya consumption does not negatively impact male hormone levels, semen quality[15] or fertility, or increase the risk of man-boobs. In fact, it actually has positive benefits. For example, it's a good source of plant-based protein and contains nutrients such as iron, calcium, folate, some B vitamins and more. Beyond that, it has also been associated with reducing the risk of prostate cancer.[16] Asian countries, which statistically have the highest rates of soya consumption, also have the lowest rates of prostate cancer, with Western countries having the highest rates. Interestingly, Asian students who move to the West and adopt a Western diet have been shown to have their risk of prostate cancer significantly increase, while Asian students who move but continue eating the diet they consumed in Asia were found to not be at a higher risk.[17]

So there's no need to fear soya, and you can even look into a mirror and say 'tofu' as many times as you like – a soy-boy man isn't going to appear.

THE SOCIAL CONFORMER

'EVERYONE AROUND ME EATS
ANIMAL PRODUCTS'

As the saying goes, if everyone around you jumped off a bridge, would you? If everyone around you went to see elephants being forced to perform in a circus, would that make animal circuses moral?

Social conformity is undeniably a significant driver of human behaviour, but the idea that we should do something because it's what the majority of people do is obviously a very dangerous way to think. The bandwagon effect can lead to all manner of atrocities and abuses, as it places the moral authority in the hands of the majority simply because they are the largest group.

The question of what is right can of course be in alignment with what the majority believe. The fact that the majority of people in the UK believe that women should have the right to vote is such an example. It is also, incidentally, an example of how societal attitudes can change and of how something that was once considered socially acceptable (the disenfranchisement of women) can become rightfully derided and thought of as being wrong.

However, what is right is not always in alignment with what the majority believe, animals suffering being one example. That said, I strongly believe that the principles of veganism already align with the values of sociey. If you asked most people if they think our food system should reduce suffering and death and be as sustainable as possible, they would say yes. That is what veganism and a plant-based diet are: an attempt to create a more sustainable and ultimately more ethical food system.

So, although most people would probably tell you that they were against veganism, if you interrogated their principles, you would find they were actually in agreement with what veganism is trying to achieve. However, getting people to line up their actions with their principles is where the problem lies, and we end up in a twisted, paradoxical system in which people do something because the majority of people are doing it, even though it goes against what they really believe. This is why it's important that we make informed and principled choices, as simply doing something because others do it can lead us into performing incredibly cruel and immoral acts. So by becoming educated citizens, we can marry our values with our actions, and encourage others to do the same too.

'IT'S LEGAL TO EAT ANIMAL PRODUCTS'

Is something moral simply because it's legal? Was slavery moral when it was legal? Is homosexuality immoral in the countries where it is illegal?

If we are to believe that morality is the same as legality, then that would mean that what is moral would change based on where we are in the world. In the UK, homosexuality would be moral because it is legal, but in Pakistan, where it is illegal, it would be immoral. History is fraught with examples of the legal system not representing what is or isn't moral, and while we would hope that what is legal aligns with what is moral, that is simply not the case, in the past or now.

Ironically, many important societal advancements that have occurred came about because of people breaking laws, and we recognise that the breaking of such laws was morally righteous, as those laws were there to uphold deeply immoral and prejudiced views and practices. It was illegal for Rosa Parks to refuse to vacate her seat on the bus but through breaking an unjust law she was fighting for something moral.

In the case of what we do to animals, our laws are a mess. Kicking animals is animal cruelty and a crime, but cutting an animal's throat and bleeding them to death is acceptable. Or at least it sometimes is. It's not acceptable to do it to a dog, or even a pig, unless of course you make money from doing it to a pig and you don't do it to just one pig but millions of pigs.

We have laws against animal cruelty and laws that protect it. Smash a car window to rescue a suffering dog on a hot day and the police will give you a pat on the back; rescue a piglet suffering on a factory farm and they'll arrest you.

In fact, it is the very worst forms of animal cruelty that take place in society that are the most legally protected. Mutilations, macerations, gas chambers, throat cutting, forced penetration, all of these practices are protected under law when done to certain animals, but kick a dog and you're an animal-abusing law-breaker. It can't just be me who thinks this doesn't make any sense.

So instead of asking whether it is legal for us to eat animal products, perhaps we should ask if it is moral for us to eat animal products.

'VEGANISM IS TOO EXTREME'

The idea that veganism is too extreme speaks to how normalised and ingrained causing harm to certain species of animals has become. When the idea of alleviating suffering is viewed as extreme, it's a pretty damning indictment of just how little worth we assign to the animals we exploit.

Why would reducing harm to others and living in the most sustainable way possible when it comes to our food choices be extreme? Surely the opposite is true. Should it not be viewed as extreme to inflict suffering on someone else, particularly when you can avoid doing so? Is it not extreme to fund places that exist for the sole purpose of slaughtering feeling, conscious beings? Are gas chambers not extreme? Isn't causing animals to scream out in pain or to shake in terror an extreme thing to do?

If two people were walking down the street and one of them started to kick a group of birds they'd come across, most of us, I'm sure, would say that the person who kicked the birds was the extreme one. Yet, our food choices are no different. We can either not participate in harming an animal or actively choose to do so. So which choice is really the extreme one?

The thing is, by labelling something as extreme, we give ourselves a reason to not engage with the issue being discussed. None of us like the idea of being an extremist, so if we label vegans as such, we create a significant barrier that allows us to disengage and simply dismiss what they are saying. This is true of a lot of the arguments used against veganism. The easiest way to avoid having to think about what vegans are advocating for is to create an image of veganism that makes it alienating and undesirable. Referring to it as extreme is one way to do that. However, in reality, veganism is the opposite of extreme.

'IT'S CULTURAL TO EAT ANIMAL PRODUCTS'

The idea that eating animals is justified because it's part of our culture is in many ways similar to the argument that eating animal products is acceptable because most of the people around you are doing it. It again subscribes to the notion that moral responsibility is diluted or even removed entirely as a consequence of the majority of people doing the same thing.

But is something moral simply because it's part of our culture? We can look all around the world right now and see why that's not the case. Homosexuality, gender equality, certain religious faiths and sects, atheism, political dissent, to name but a few, are all examples of things that are either frowned upon, illegal or even punishable by death in many cultures.

The inverse is true as well – while something can be moral yet culturally unacceptable, something can be immoral yet culturally accepted. What we do to animals is an example of this. Eating animal products is culturally accepted more or less everywhere, but that doesn't in and of itself mean that it is moral as a result.

We might feel that cultural practices like the consumption of dogs, cats, whales and dolphins are immoral, but they are immoral not because we don't consume those animals in our culture but because consuming those animals means violating their autonomy, causing them suffering and taking their lives from them. It is for these same reasons that the consumption of pigs, cows, chickens, lambs, fish and the other animals we consume is also immoral. Consuming a dog is not morally different to consuming a pig. The only difference is our perception, which is based on our cultural relationship with dogs.

And culture is simply a behaviour that a group of people have been carrying out for a period of time. Culture also changes and evolves. And thank goodness it does, because it wasn't that long ago that in the UK women didn't have the right to vote, Black people could be owned as slaves, and blasphemers could be sentenced to death.

So what's more important, upholding something simply because it's cultural or reducing suffering, exploitation, harm and death?

'VEGANS HAVE HIGH RECIDIVISM RATES'

Before we look at the study that is always used to illustrate this argument, it's important to recognise that the recidivism rates of something don't prove whether or not that thing has merits or is indeed preferable. Take exercise as an example – the majority of people who start a gym membership end up not actually using it. Does that mean that exercising is bad for us?

Recidivism rates and the reasons for them are important in determining what difficulties people face and how to make it easier for them to stick at something. However, they don't prove that something is inherently good or bad. So even if the majority of people who go vegan don't end up staying vegan, that doesn't invalidate veganism or its merits – it just shows that there are barriers to retention that need to be addressed and improved.

However, recidivism rates are often used to suggest that veganism is unhealthy and that's why people don't continue with it. As Joe Rogan put it on his podcast, 'Most [vegans] will quit. They're going to quit veganism and they're going to start eating meat again because of their health. That's the truth. It's some ridiculous number.'

So what is the study that is most often cited, and does it support the claim that Rogan and others like him make?

A group called the Humane Research Council (and now known as Faunalytics) surveyed former and current vegetarians and vegans and found that 84 per cent abandoned their diets.[1] This figure is the one that is now regularly touted.

However, there are a couple of significant issues with the study. First, it was published in 2014, which might not seem that long ago but is actually a very long time when it comes to veganism. I wasn't vegan in 2014, and the majority of vegans I know or have spoken to also weren't vegan back then. This is important, because the biggest reasons cited for readoption were social and practical issues.

Since 2014, veganism has become more mainstream, more socially accepted and more convenient, and plant-based alternatives are now tastier and more varied. In other words, many of the biggest barriers former vegetarians and vegans faced have actually become less significant since the survey. There is also a lot more awareness today about the issues related to animal farming, so it is now easier to find information that incentivises and motivates people to make the change. Plus, those surveyed might have stopped being vegetarian or vegan before 2014, meaning the changes for them could be even more significant now.

It's also worth saying that lumping vegetarians and vegans together actually made the recidivism rates for vegans look worse than they were. While 86 per cent of surveyed vegetarians abandoned their diet, 70 per cent of surveyed vegans did. Only 27 per cent of the former vegetarians and vegans listed animal protection as one of their motivations for changing in the first place, and only 23 per cent listed concern for the environment. The biggest initial motivation for former vegetarians and vegans changing their diets was health. This is an indicator that those who treat veganism as a health-based diet are more likely to stop being vegan than those who treat it as a lifestyle that can be used to achieve some form of social good. When viewed in these terms, the rates might actually reflect well on plant-based diets, considering the rate of recidivism for all diets is estimated to be 97 per cent.[2]

Interestingly, the majority of those surveyed who quit

being vegetarian or vegan did so within the first 3 months, and then the next largest drop-off was between 4 and 11 months. In fact, more than 50 per cent of people who went vegetarian or vegan and then stopped did so within a year. However, of those surveyed who listed themselves as currently vegetarian or vegan, the largest group (58 per cent) were those who had been vegetarian or vegan for more than 10 years.

If it were true that veganism is unhealthy in the long term and causes health complications, you would expect to see a jump in the number of people who gave up being vegan over time, but the percentage of those surveyed who stopped being vegetarian or vegan grew smaller with each time increment surveyed and then levelled off after six years. In fact, more than three-quarters of the former vegetarians and vegans said they had no concerns about the impact their diets were having on their health, which is in direct contradiction to the claims that Joe Rogan and others who use this argument make.

So while we don't know the true recidivism rates of vegans, especially in more recent years, the main take-home message from the study that is cited to highlight this criticism of vegans is that people who view veganism as a diet are less likely to stick with it, as is the case with people who adopt diets in general. Beyond that, the recidivism rates related to veganism are ultimately irrelevant in ascertaining its true merits.

THE 'WHAT-ABOUTER'

'WHAT ABOUT THE UNETHICAL INDUSTRIES THAT VEGANS PARTAKE IN?'

It is true that vegans can partake in industries that are not wholly ethical. However, this argument is an appeal to futility because it is impossible to be perfect.

Should we be looking to travel in more sustainable ways? Yes, absolutely. Should we try to avoid fast fashion? Yes, absolutely. Should we cut down on our food waste? Yes, absolutely.

But none of these are arguments against veganism. They are simply arguments in favour of thinking about our impact on the planet and on others beyond veganism. Plus, the systems that we exist within are not systems that we directly chose. If we had a blank canvas, would we decide to draw the same version of the world that we currently live in? Of course not.

However, one of the empowering aspects of veganism is that it is one of the few opportunities we have of making a genuinely impactful choice. Let's say in your home you have two kinds of electrical sockets: one is powered by fossil fuels and the other by renewable energy. Which socket is the right one to plug into? At the same time, let's say that you are unable to be vegan; does that mean that you shouldn't use the renewable energy socket?

The consumer choices we face are the opposite of that scenario. We do have the choice to be vegan, but we don't have a choice with many of the other problem industries that we partake in. Does that mean that we shouldn't be vegan because we can't make completely perfect choices in other regards?

Being vegan doesn't provide a free pass to not think about other areas of harm. However, the fact that vegans aren't perfect doesn't provide a justification to not be vegan either. Simply put, veganism on its own won't bring about an entirely ethical and equitable world, but, at the same time, a truly ethical and equitable world won't exist without it.

'WHAT ABOUT THE BEES THAT ARE USED TO POLLINATE CERTAIN CROPS?'

This anti-vegan argument became popular after a 2018 episode of the quiz show *QI*. In the programme the host, Sandi Toksvig, asked the panellists which of the plant-foods they were being shown weren't vegan, before telling them that actually none of them were. The reason being that 'they can't exist without bees, and the bees are used in, let's call it, an "unnatural way". Because they are so difficult to cultivate naturally, all of these crops rely on bees which are placed on the backs of trucks and taken very long distances across the country. It's migratory beekeeping and it's unnatural use of animals and there are lots of [plant]-foods that fall foul of this.'

So, let's break this down.

The fact that foods are pollinated is, of course, not in and of itself a problem. I mention this only because sometimes the logic of this argument is extended to say that bee-pollinated plant-foods aren't vegan because they rely on animals for pollination. But pollination is not an inherently exploitative or harmful event. In fact, it's essential for the pollinators and beneficial for them. So the issue at hand isn't that insects pollinate plants, it's the manner in which that process sometimes occurs. For example, the trucking of hives to farms.

Importantly, one of the reasons why this practice of migratory beekeeping occurs is due to the sheer amount of food that is produced, leading to farming practices that are unable to rely on more traditional or natural forms of pollination. Crops grown in pollinator-friendly environments obviously allow for a more symbiotic relationship to occur. It is also worth pointing out that these farms are not run by vegans, and the methods and practices used on them were not devised by vegans or with the best interests of animals in mind. A huge part of the reason why we have issues like this one is because we have created the system we have by viewing animals as not deserving of moral consideration.

No wonder we use migratory beekeeping when our agricultural industry views animals as disposable. Yet it is vegans who are called hypocrites by non-vegans who not only eat animal products but also eat almonds, avocados, cherries, broccoli and the other foods that can be pollinated by honeybees. Plus, the main reason why it is honeybees that are used in this process is because honey producers rent them out to crop farmers to make extra money on the side. It's cheaper and more convenient to do this than to have their own permanent hives and incur those costs. It is therefore because the honey industry exists that these practices occur like they do. If the honey industry didn't exist, and beekeepers weren't renting out their bees, crop producers reliant on pollinators would have to adapt.

One of the ways they could do this would be to have permanent hives themselves, reducing the concerns around bee transportation and spread of diseases between pollinator populations, although this would not be the solution I would advocate for. Alternatively, and preferably, farmers would adopt practices that encouraged wild pollinators.

Honeybees are just one species of bee, and bees are not the only pollinators that exist. We use honeybees because we already have billions of them that are being rented out specifically for this use.

By not consuming honey, vegans are actually taking a stand against the use of honeybees for pollination, but it's simply not possible to remove our participation from these practices completely, as we don't know exactly where they happen or which products have been pollinated in this way.

However, a reduction in the demand for honey would have a positive impact. A UK study into honeybee use stated:

> In the early 1980s honeybees provided most of our pollination services, however, following severe declines in hive numbers over the last 30 years, there are no longer enough honeybees to do the job and it is now wild insects, such as bumblebees and hoverflies, that have filled the void to ensure that our crops are pollinated and our food production is secure.[1]

Honeybees only pollinate between 5 per cent and 15 per cent of the UK's insect-pollinated crops, with wild bees pollinating 85–95 per cent.[2] In other words, reducing the number of honeybees incentivises different forms of pollination.

Another important aspect of this debate is the fact that not all crops pollinated by honeybees are for human consumption, as primary feed crops are pollinated by honeybees too, including for dairy cows. Considering this argument about bee pollination most often comes up with regard to vegan consumption of almond milk, it's certainly ironic that dairy cows are given feed that comes from crops that can be pollinated with honeybees. And the solution to the problem of migratory bee pollination of almonds, particularly in California, isn't to reject veganism, it's to get almonds from places like Europe, where pollination is largely dependent on wild bees and wild pollinators. Or you can drink oat milk, as oats self-pollinate and don't require bees at all.

Maybe I'm cynical, but I can't help but feel that the reason why we only hear about bee welfare – when they're being weaponised as an argument against veganism – is because we're more interested in trying to justify not being vegan than we are genuinely concerned about these animals. Because if we really cared about bees, instead of supporting an industry that uses more land, contributes to more habitat loss and also increases the amount of cropland we need, we would all be vegan, in turn reducing the amount of plants being grown and pesticides being used, and also then allowing us the ability to re-wild huge areas of land that we could use to increase flora and in turn improve biodiversity, including the numbers of wild pollinators like bees.

In fact, veganism is the best thing we can do for bees, because not only does it reduce demand for honey, thereby significantly limiting the use of migratory honeybee pollination, but it also provides us with the most effective way of restoring populations of wild bees and other pollinators.

And the message behind this argument is so self-defeating. It's basically saying, if you try to make a positive difference you can be criticised for not being perfect, but if you don't try then you can engage in those same systems and also avoid any scrutiny or criticism at the same time.

Simply put, just because the plant-based food system we have right now isn't perfect isn't a reason to continue supporting a system which is far worse, especially as doing so further perpetuates the problems found within plant-based agriculture. Instead, let's be vegan and also work towards creating systems of plant production that are more sustainable, more harmonious with the natural world and, as a result, better for wild pollinators and insects as well.

'WHAT ABOUT PLANTS?'

Checkmate, vegans. You eat plants and plants are alive, so why don't you care about their feelings?

Let's say your house is on fire and you've escaped but your dog is still inside. A fireman rushes in and comes out a couple of minutes later holding the plant you had on your windowsill. Would you say, 'Thanks so much for saving the plant. I was equally upset about the plant and my dog'? Or let's say you're driving home and a cat runs out in front of your car. You have two options: run over the cat or swerve onto a bed of roses. Which do you choose?

If we are to believe that animals and plants are morally the same, we need to establish why, and we need to consider the implications of what that would mean in practice. In terms of the latter, to say that plants are morally the same as animals would mean making the harming of plants the same as the harming of animals (humans included) and as such we'd be creating a moral equivalence between mowing the lawn and mass killing. A hobby gardener would be the moral equivalent of a serial cat killer. If we don't believe that these two actions are the same, we have already established there is a distinction between doing something to a plant and doing something to an animal.

As for why some people say plants and animals are morally the same, the reasoning tends to come down to one of two things: either because they are alive or because of their intelligence. Let's first look at the argument that plants and animals are morally the same because they're both alive, a point that the American astrophysicist Neil deGrasse Tyson has made on his YouTube channel, to defend his consumption of animals: 'We basically eradicated smallpox, right? Well, what about the smallpox microbes? How do they feel about this?'

Despite not mentioning plants specifically, by questioning the morality of eliminating microbes such as smallpox he's taking the argument to its logical conclusion and trying to make vegans seem like hypocrites. What he actually highlights is just how absurd this argument really is. Because if we are to claim that all life is equally valuable simply by virtue of being alive, the conclusion is that microbes are as morally relevant as humans. This would make taking antibiotics an immoral act, as they kill millions of microbes. In fact, if all life is morally equivalent, then taking antibiotics would be one of the most immoral things you could do because of the number of living microbes you are killing.

However, if you don't believe that bacteria possess intrinsic moral worth even though they are a form of life, then you've already ascertained that not all life is worthy of equal moral consideration. Once we've come to this realisation, we then have to work out why and decide what the meaningful criteria for assigning moral worth to living organisms is. Which brings us on to intelligence.

Plants are undeniably capable of doing some incredible things. The way they grow and interact with their environments is impressive and fascinating, and could by some definitions be described as a sign of intelligent behaviour. However, intelligence does not equal sentience or the capacity to experience subjectively. Going back to microbes, they too display forms of intelligence and are capable of doing remarkable things. They will mutate and adapt in order to survive, which is very impressive. But that doesn't mean that microbes are sentient beings with subjective experiences.

Someone arguing that plants are sentient might point out that they communicate with one another. For example, some secrete chemicals to send signals to other plants. But bacteria also communicate with one another using chemical signal molecules that allow them to 'monitor the environment for other bacteria and to alter behavior on a population-wide scale in response to changes in the number and/or species present in a community'.[3] So is taking antibiotics morally comparable to harming animals? If not, then why is eating plants?

Plants have predetermined responses to certain stimuli. For example, the reason a Venus fly trap closes around a fly is not because it's had a conscious reaction and has decided to close around the fly – it's because the fly has triggered the stimulus causing the plant's predetermined response to occur as a result. This is why anything that triggers that response will cause the same outcome, cigarette butts being an example. A pig or a cow, on the other hand, won't just eat a cigarette butt because you offer it to them as food. They have a conscious reaction, which is to not eat it because they don't want to.

The word 'want' is really important. Animals do things because they want to. A cat who walks over to a mug on the side of a table and decides to knock it over the edge so it smashes on the floor does so because they want to, not because they have a predetermined response to be a cute yet calculating menace.

From a purely anatomical perspective, plants don't have a brain, nervous system or pain receptors. Furthermore, from an evolutionary perspective, the reason why humans and non-human animals feel pain is so that we can safeguard ourselves from harm, remove ourselves from dangerous situations, and learn what things to avoid doing. The capacity to feel pain or have subjective experiences is completely useless to a plant; in fact, it would be torturous and tormenting. So not only do they lack the biological functions needed in order to have feelings and to be sentient, but there's no evolutionary reason why they would have these traits either.

One of the most telling things about this argument is that it is only ever used when people are trying to justify causing needless harm to animals. Have you ever heard someone who murdered a human say in their defence, 'Your Honour, I did kill that man. But have you ever picked the leaves off a daisy?'

In many ways this argument speaks to just how little we value other animals. The fact that we would never make such an argument to defend harming humans shows not only how inconsistently and disingenuously we use this logic, but also speaks to how far we have degraded and demeaned the animals we consume. The only way we could sincerely believe that eating a cucumber is morally the same as forcing a pig into a gas chamber is if we view pigs as being practically worthless.

Even if we ignore everything I've just outlined and run with the logic that plants and animals are the same, this is actually an argument for veganism, because although going vegan would mean that as individuals our plant consumption would increase, it would actually mean fewer plants being consumed overall. This is because the animals we eat also eat plants. Then of course there are all the grazing animals who spend their days tearing up, crushing and eating all that poor grass. Every day in a cattle field is a hellish experience for all that defenceless grass just waiting for it to be its turn to be tortured and killed by the cattle. And on top of all that, animal farming is the number one driver of deforestation and habitat destruction. So animal farming kills plants to create space for animals and then kills more plants to feed them before ultimately killing the animals as well. Simply put, if you care about plants and want to minimise the harm caused to them, then you have a moral obligation to be a vegan.

'WHAT ABOUT THE ANIMALS THAT VEGANS KILL?'

I count myself lucky that I managed to spend a good portion of my life unaware of the thoughts and opinions of Ted Nugent. However, all good things come to an end at some point.

Ted Nugent: 'If you really want to kill the most things, be a vegan, because the farmers who protect your beans kill everything.'[4]

This argument, commonly known as the crop-deaths argument, states that vegans are responsible for the death of more animals because of crop production than those who consume grazing animals. But where does this idea come from?

Back in 2003, an animal science professor by the name of Steven Davis published a piece in which he sought to argue that an omnivorous diet that included pasture-raised animals was preferable to a solely plant-based one.[5] Davis did this by calculating that 7.5 animals are killed per hectare of ruminant pasture and 15 are killed per hectare of land for growing crops, with these deaths usually consisting of small animals like rodents being killed by the harvesting itself or by predators in the surrounding areas.

Even if we were to accept these estimates as being true, he makes a fundamental mistake by assuming that equal amounts of food are produced per hectare of land, which is clearly not the case. For example, data from the UN states that 1,000kg of plant protein can be produced on 1 hectare of land (although yields can be higher), but it would take 10 hectares of land for grass-fed beef to produce the same amount.[6] Using Davis's estimates and combining them with the amount of land needed to produce an equal amount of protein shows that vegans are in fact responsible for five times fewer animal deaths.

One of the studies that Davis used to come up with his estimates involved fitting 33 field mice with radio collars and tracking them before and after a harvest. While 55 per cent of them died following the harvest, only 3 per cent (or one mouse) were killed by the actual harvesting, with the other mice being killed by predators.

Although it could be argued that the loss of crop cover contributed to the death of the animals who were killed by the predators, it's not as if those predators wouldn't have eaten without the crop harvest. They might not have eaten the specific animals that they did, but an animal would still have died because that's how those predators eat. So, are these deaths because of crop production or because predators hunt and kill prey to survive?

Another study that examined the effect of wheat and corn harvesting in central Argentina compared the population and distribution of grass mice in three habitats: crop fields, regions bordering the fields and the wider surrounding area. While the number of mice found in fields substantially decreased after harvest, their numbers substantially increased in the border regions. When it came to 'disappearances' – a category that included both mouse deaths and migration out of the study area – there was no significant difference between the three habitats. The study concluded that changes in the number of field animals were 'the consequence of movement and not of high[er] mortality in crops'.[7]

Davis's methodology also failed to factor in other considerations, such as the branding, disbudding, dehorning, castration and ear tagging that can occur in pasture-based systems, nor did it account for the suffering and distress caused to animals when they are loaded into trucks and taken to slaughterhouses. However, even though his study was hugely flawed, this piece of research has persisted and is still quoted today.

Not to be outdone, another article was published back in 2011 that was arguably even worse but which has since been repeated as gospel by certain corners of the anti-vegan world. The article, written by Mike Archer, claimed that wheat production is responsible for 25 times more deaths than grass-fed beef.[8] Why? Because in Australia every four years on average there are events called mouse plagues where an overwhelming number of mice overrun the fields and are often poisoned as a result.

This argument boils down to it being bad for me to be a vegan in the UK because there are mouse plagues in Australia (and occasionally China too) – how does that make any sense? It's not as though vegans only eat plants grown in Australia! This argument is even flimsier when you consider that the mice plagues don't just impact plants grown for humans – they also affect crops grown for animal feed, something the author astonishingly didn't factor into his calculations. This is especially egregious when you consider that approximately 1.7 times more wheat was used as animal feed than food for humans in Australia in 2019–20.[9]

Even solely grass-fed cattle can still be given things like sorghum, hay and silage made from grass. These plants are also harvested and therefore affected by the mouse plagues. And it doesn't end there – mouse plagues also destroy pasture and impact grazing land. Again, none of this was factored into the author's calculations.

If you thought all of that was bad enough, the numbers that Archer used in his article were incorrect. He overstated the scale of the mouse plagues by calculating that each area of grain production in Australia has one every four years; however, this is not the case. A report looking into not only

Archer's numbers but also the wider argument about crop deaths recalculated the numbers and stated:

> A more accurate picture is suggested by the Cooperative Research Centre which notes that each year between 100,000 and 500,000 hectares of grain crops in Australia are subject to mouse plagues. These figures suggest that in an average year 2.3 per cent of Australian grain cropland is hit by plague.[10]

Archer's figures calculated that 2.2 animals are killed per 100kg of usable grass-fed beef protein and 55 are killed per 100kg of plant protein. When these are updated to reflect the actual scope of the plagues, 1.27 animals are killed per 100kg of usable plant protein. And it is also worth bearing in mind that the 2.2 animal deaths per 100kg of usable grass-fed beef protein does not include the animals killed for the harvest of hay, silage and other feed, or for the protection of pasture, so that number will actually be higher as well. It is absolutely outrageous that such an awful piece of research is used in an attempt to debunk veganism. The authors of the review of the evidence regarding crop deaths summed it up nicely when they stated, 'To date, Steven Davis and Michael Archer have offered the most extensive empirical information about animal deaths in plant agriculture – which, as will soon become apparent, isn't saying much.'[11]

Finally, they concluded, 'Agriculture has taken a wide variety of forms throughout history, and current trends would seem to raise the serious possibility that plant agriculture might someday kill very few animals – perhaps even none.'

Once we establish that veganism contributes to fewer animal deaths than eating animal products, the crop-deaths argument becomes an example of the Nirvana fallacy, whereby realistic ideas and solutions are held up to an impossible standard as a means to disingenuously discount them. In the case of veganism, its merits are judged against an unrealistic expectation of it being completely without fault or harm. Of course animals can die for plant-foods as well. Of course a plant-based diet isn't perfect. However, these are not the claims that vegans make. The argument isn't that veganism is faultless but instead that veganism is a better choice.

THE COMPROMISER

'INSECT FARMING IS A SUSTAINABLE ALTERNATIVE TO MEAT AND DAIRY'

Insect farming is increasingly touted as an alternative to conventional animal farming, with certain supermarkets even stocking a small selection of insect-based foods. Proponents argue that one of the major benefits to insect farming is its low environmental impact. But is insect farming really more sustainable than current animal farming?

Based on current data, it certainly seems to be. Insect farming uses less land, less water and emits fewer greenhouse gasses than conventional animal farming,[1] especially in the case of red meat. With that said, a group of scientists published an article in 2019, stating that 'a lack of basic research on almost all aspects of production means the future environmental impact of the mass rearing of insects is largely unknown'.[2] It appears as though there would be environmental benefits to an insect-based system compared to what we have now, but the extent of those benefits and indeed the impact of large-scale insect farming is something that needs further exploration.

There is, however, one clear benefit to insect farming: the feed conversion ratio. This refers to the amount of feed needed to produce a certain amount of food or nutrients from the animals being fed. In the case of conventional animal products, the most protein-efficient foodstuff is eggs;

however, they're only 25 per cent efficient, meaning that 75 per cent of the protein is lost during the conversion. For beef, the conversion is 3.8 per cent, meaning that more than 96 per cent of the protein in the animal feed given to cattle is lost by the time it gets to us. In terms of calories rather than protein, the most efficient foodstuff is whole milk at 24 per cent, and again the worst is beef at only 1.9 per cent.[3]

When it comes to insects, while the numbers vary, they are generally considered to be more protein and energy efficient than conventional animal products, with it estimated that crickets could be 12 times as efficient as cattle when it comes to converting feed into meat.[4] However, the real question is not how insects stack up against cows, pigs and chickens, but how they stack up against plants. Farming animals always creates inefficiency, as it requires the conversion of plant matter into animal matter, a point backed up by Dennis Oonincx, an expert in edible insects and sustainability: 'Plants that can be consumed directly are best used as food instead of feed for insects.'[5]

An analysis looking at protein and energy efficiency found that while mealworms and crickets were more efficient than other animal products, they were less efficient than soybeans, which had the highest energy and protein yields of the foods analysed.[6] The lead author even said that eating bean burgers was the most sustainable option available to us based on the foods they analysed.[7]

Another argument for insect farming is that insects can be raised on food and crop waste, resulting in increased sustainability. But it isn't quite as simple as this. First, the waste can be used for other purposes, such as to increase soil health via the production of compost and farm fertiliser. Second, food waste should be reduced, not encouraged. If an industry becomes reliant on waste products, how does it expand without more waste? Ultimately, reducing food waste is a better option than finding a way to commercialise it. And third, food waste is not the optimal feed for insects anyway. A study found that crickets given poultry feed ended up 75 per cent larger in weight than those being fed food waste, plus they gained that larger weight six days faster.[8] So the idea of food waste is a red herring. Finally, scientists are already creating genetically modified versions of the insects we currently farm, exploring how they can be modified to grow larger and in less time. But what if these genetically modified insects escaped? How would that impact other insects and what would be the negative consequences of such an event? It could be catastrophic.

So, if growing plants is more efficient than insect farming, more sustainable and less likely to lead to biodiversity disaster, it makes much more sense to create a circular system where the waste from plant production is put back into producing more plants for us to eat.

'INSECT FARMING IS AN ETHICAL ALTERNATIVE TO MEAT AND DAIRY'

If eating insects is not more sustainable than just eating plants, is it more ethical?

Insects are commercially farmed in plastic bins or trays, where they'll be kept their entire lives until they reach slaughter weight. They are then usually killed by either being packed together and then frozen into a block, or by being ground up and turned into a powder. Alternatively, some companies will steam, boil or suffocate the animals to death.

Whether or not this is ethical treatment again comes down to the insects' sentience and their capacity to suffer and experience pain. Knowing what we already know about animal sentience, we could take the position that eating insects is morally justified because they are less sentient than animals such as pigs, lambs, cows, chickens, fish and the other animals we conventionally eat. But the question shouldn't be: are they as sentient as other farmed animals? It should instead be: are they more sentient than plants?

Insects have a nervous system and nociceptors – often referred to as pain receptors. Plants don't have these. Some insects also produce endorphins, which are hormones that are released when an animal feels things such as pain and stress. Insects have been shown to have the ability to learn, such as avoiding stimuli that they associate with negative experiences[9], or actively choosing certain stimuli that they associate with a reward, such as food.[10] Experiments have been done in which fruit flies have been seen to act in a way that suggests they experience chronic pain.[11] And crickets have been shown to react to receiving morphine, staying in a box that was getting progressively hotter for a longer period than the crickets who were not given the drug. After five days of being given morphine, they even started to exhibit signs of addiction.[12]

It's still early days when it comes to our understanding of insect sentience, but we already have strong evidence to suggest that we should view the way we treat them as an issue of moral concern. This is especially the case when we consider the sheer number of insects we're talking about. A current estimate places the number of insects killed for either animal feed or human food at around 1.2 trillion. On top of this, a further 2 trillion wild-caught insects are estimated to be killed for human food each year as well.[13] And to get the same amount of protein from crickets as we do from one cow would require us to kill somewhere in the region of 350,000 crickets.[14] So even if we view the moral worth of a cricket as being less than that of a cow, the choice isn't between killing one cow or one cricket – it's one cow or 350,000 crickets.

A plant-based diet isn't perfect and does also result in insects being killed, but although we can't currently have a diet free from animal harm, eating plants over animals (insects included) is still the diet that reduces animal suffering and death overall.

And while it's true that insects are consumed all around the world, with hundreds of millions of people eating them, there is still the question of perceived edibility in cultures where it doesn't have a historical precedence. We could call this the 'yuck' factor. It's already hard enough trying to get people to stop eating animals when they're being offered foods that they are culturally used to. If trying to get people to eat lentil and mushroom burgers is difficult, I'm not sure if a crushed black soldier fly larvae burger is going to be an easier sell, let alone a more ethical and sustainable one.

If there is a criticism that plant-based diets are still not accessible enough, then insect-based diets are even less accessible. In fact, insect-based foods are still more or less non-existent in Western countries, especially to the extent where they could act as an actual replacement for people to stop eating the conventional animal products they consume. So does it really make sense to invest time, money, resources and infrastructure into a new model of farming when we already have an alternative to our current model of animal farming? Plant farming is not only considerably more accessible and affordable, it doesn't present the same 'yuck' factor challenges, and, most importantly, is more ethical and sustainable than insect farming anyway.

'HONEY DOESN'T HARM BEES AND IS GOOD FOR THE ENVIRONMENT'

Following on from the broader insect discussion, let's turn our focus to one insect in particular: the honeybee. Some people argue that not including honey in a plant-based diet makes veganism seem more extreme and inaccessible. However, to understand whether honey should be seen as a compromise we have to deal with two particular questions: Is human consumption of honey really bad for bees? And isn't having more honeybees actually a good thing?

In regards to the sentience of honeybees, there is strong evidence that they can demonstrate emotions and even exhibit pessimism and optimism.[15] They've also been shown to memorise human faces,[16] learn to use tools and solve problems.[17] Importantly, research has shown that they modify their behaviours when faced with scenarios that are noxious or indicative of causing pain.[18] The lead researcher of one study stated that bees act in a way that 'is consistent with the capacity of a subjective experience of pain'.[19]

So with the growing evidence around the complexity of bees, where does that leave us when it comes to honey? Conventional beekeeping involves several unethical practices. These include instrumental insemination, where typically between 8 to 12 drones are crushed to death and have their semen extracted from them. The queen bee is then restrained and has the semen injected inside of her.

To help beekeepers identify the queen bee or to prevent swarming – this is when a single bee colony will split into two or more distinct colonies, which is bad for business as it reduces the honey production from that hive – it is common for the queen bee to have either one or both of her wings clipped. You can actually buy queen bees online that have had their wings clipped.

Honey is a bee's food. They produce it by swallowing nectar, regurgitating it and then repeating this process many times. It takes about 12 worker bees an entire lifetime to create a single teaspoon of honey. Beekeepers normally take the honey from the hive during autumn. To sustain them through the winter months, the bees will instead be given a sugar syrup, which is an inadequate replacement. This, coupled with the fact that honeybees are selectively bred and have a narrowed population gene pool, can lead to diseases and large die-offs. Pathogens that affect honeybees, like Varroa mite and deformed wing virus, can also spread into wild bee species.

Instead of using a replacement syrup, some beekeepers will kill off the hives for the winter months and start again in spring, as that can be cheaper. Ways that beekeepers kill bee colonies include pouring petrol into the hive, drowning them with soapy water, suffocating them in large industrial bin bags or gassing them to death.

But what about smaller-scale beekeepers who don't kill off their bees, or even hobby beekeepers? Well, even hobby beekeepers need to get their bees from breeders, which means supporting the insemination and transportation of these bees. So even small-scale beekeeping involves some of the same practices as commercial beekeeping.

But there are other problems. It's no secret that we are in the midst of a pollinator crisis, with bee populations declining. The question then becomes: isn't eating honey a good thing because we need more bees? The Department of Zoology at the University of Cambridge doesn't think so: 'The crisis in global pollinator decline has been associated with one species above all, the western honeybee.'[20]

Let me explain. There are approximately 20,000 species of bees worldwide – the UK alone has 270 different bee species. The honeybee is one species that we have introduced by the billions. So what impact does that have on wild pollinators? The Canary Islands provided the perfect place for researchers to find out. A study was done looking at how the introduction of honeybees affected the native wild pollinators. Sadly, the introduction of the honeybees created a disruption between plants and pollinators, and there was a decline in ecosystem resilience.[21]

Other studies have shown that honeybees actually harm wild bee populations because they can outcompete wild bees when it comes to collecting nectar and pollen.[22] While they can be more dominant, honeybees are also less effective at pollination, making it harder for wild plants to reproduce.[23]

Sheila Colla, an assistant professor and conservation biologist at Toronto's York University, put it succinctly when she stated, 'Beekeeping is for people; it's not a conservation practice.' And experts from Kew Gardens have also sounded the alarm, stating: 'Campaigns encouraging people to save bees have resulted in an unsustainable proliferation in urban beekeeping. This approach only saves one species of bee, the honeybee, with no regard for how honeybees interact with other, native species.'[24]

'IT'S OK TO EAT BIVALVES LIKE MUSSELS AND OYSTERS'

The question of whether or not it is vegan or ethical to consume bivalves is actually a very interesting one. Bivalves are aquatic animals that are so named because of their two hinged shells. Bivalves include mussels, oysters and clams, and it's estimated that we consume somewhere in the region of 15 million tonnes a year, with around 90 per cent of those coming from farms.[25]

The justification for vegans eating bivalves stems from the theory that without a brain or central nervous system, there is a good chance that they are not sentient and lack the required biological functions to feel and experience pain. The argument, therefore, is that there is no significant moral distinction between consuming them or plants.

Sometimes it's argued that eating bivalves is wrong simply

because they're animals, but the reason vegans don't eat animals is not because they are animals but because they are sentient and can feel. Using the argument 'because they're an animal' is akin to someone saying humans are superior 'because they are human'. There needs to be something that qualifies what it is about being an animal that makes their life and experiences important to consider. So, if sentience is central to whether or not eating an animal is ethical, the question is: are bivalves sentient in some capacity?

Very little research into the sentience of bivalves has been done. That being said, the literature that does exist points to some interesting findings. While bivalves don't have brains, they do have nerves and ganglia, which are basically clusters of neurons. Yes, their anatomies might not be as complex as those of other animals, but these attributes separate them from the plant kingdom.

In one study, mussels were shown to change their behaviour in order to avoid parasites they had previously encountered, even though this change in behaviour came with trade-offs, such as reduced energy uptake and growth.[26] In another study, mussels were shown to reduce their defensive behaviours after repeated exposure to the same sounds, suggesting that they are able to alter their responses according to perceived danger levels.[27] They've also been shown to have an aversive response to cold water[28] and release morphine when their muscles are damaged.[29] Scallops and clams also evade danger by swimming away from predators by clapping their shells.

However, these findings don't prove that bivalves are sentient, nor do they prove that they feel pain. But this lack of research doesn't necessarily give us a good reason to keep eating them. The fact that we don't know whether they are sentient or feel pain means we should tread carefully and give bivalves the benefit of the doubt. After all, if in the future we discovered they weren't sentient, the worst thing we would have done is deprive ourselves of a few portions of moules marinières. As it stands, it seems safe to say that although eating bivalves is morally preferable to eating other animals, such as cows and pigs, there is no real reason to incorporate bivalves into our diet, as we can just eat plants.

There are also some environmental concerns around bivalve fishing, including dredging, which is similar in technique to bottom trawling and can cause significant environmental degradation and the deaths of animals caught as by-catch.

However, the argument about eating bivalves is complicated even further when we factor in the potential suffering of eating bivalves versus the potential suffering caused to animals in the production of plants, particularly insects. If it was simply the case that the plants we ate didn't cause any animal harm, then the decision to eat bivalves or plants could be made on the basis that the capacity for bivalves to experience is greater than that of plants. However, given that our current plant-based farming system is not without animal harm, how do we then quantify the relative harms caused by eating bivalves versus eating plants? There are no definitive answers, except that whichever way you fall doesn't provide a justification for eating any other type of animal product.

'WE CAN COMPROMISE AND JUST REDUCE THE AMOUNT OF ANIMAL PRODUCTS WE EAT'

At face value, this argument has obvious merits – by reducing the number of animal products we eat, we can reduce the impact of animal farming. In addition, the proposition of simply reducing rather than entirely abstaining from animal consumption is, at least for a lot of people, a more appealing one. It almost presents a best-of-both-worlds scenario, a compromise that satisfies everyone. Except, does it?

From an environmental perspective, reduction is in many ways the name of the game. We can't eliminate all of our negative impact on the planet, so in essence our aim is always simply to reduce it. Plus, many of the issues we face are due to scale: one pick-up truck is not going to significantly impact the planet; hundreds of thousands, however, is a different story entirely.

But just because humans will always have some impact doesn't mean that we shouldn't aim to minimise that impact as far as possible. In the case of diet, eating plants is more sustainable, so that should be the default from an environmental perspective.

However, it is the ethical dimension that really highlights the flaws in the reduced-consumption argument. To begin with, the issue of animal exploitation is not simply one of scale, although it is clearly true that the more individuals you cause harm to the worse your action becomes. The fact that we slaughter at least 85 billion land animals a year and as many as 2.3 trillion marine animals is simply an indictment of how wildly monstrous our use and abuse of animals has become. But these huge numbers actually provide a smokescreen from the real issue at hand. This issue is not simply one of scale. Animals are individuals and each of them suffers and experiences as an individual.

The animal cruelty laws we have don't make abusing one dog permissible but five dogs a crime. We also wouldn't look to compromise with someone who wanted to kill ten dogs by advocating that they should only kill five. Simply put, for those five who would still be killed that compromise means nothing.

And herein lies the issue. While reduction is great news for the animals who will be spared, for those who are still being exploited it means nothing. The suffering of a pig in a gas chamber is not alleviated because fewer other pigs are also going to be killed.

However, let me add an important caveat to this. I can appreciate that for any readers who are not vegan (and maybe even some who are) the rationale in this section might seem dogmatic and alienating. For that reason, whenever someone tells me that they have cut down or reduced their animal-product consumption, my first response is to validate the effort that they have put in and acknowledge the good intentions behind their choices so far. I would then hopefully empower the person to recognise that by eliminating the remaining number of animal products they consume, they would be one step closer to achieving the aims that motivated them to reduce in the first place.

If I promote something less than veganism, the potential for someone to go vegan as a result of talking to me decreases. After all, why would someone go vegan if I'm not presenting an argument for them to do so? However, advocating for veganism means people will be more likely to go vegan and at the same time it won't dissuade people who are planning on reducing their consumption from doing so.

I also believe that the rationale behind promoting reduction because it's an easier sell is ultimately patronising. It implies that people are incapable of making tough decisions or grappling with challenging concepts and should instead be given something easier to deal with. To me, that's condescending, especially as I'm someone who has gone vegan. It would be like saying that I'm able to do something that I don't believe others will be able to do.

I also fully acknowledge that many people won't go vegan overnight. However, the argument that I am making is that a reduction shouldn't be the end goal.

'I ONLY BUY WELFARE-ACCREDITED ANIMAL PRODUCTS'

Improving animal welfare is sometimes seen as a compromise in that it appeases the desire of vegans to help animals while also appeasing the desire of non-vegans to eat animal products. But does animal welfare actually tackle the fundamental problem?

Animal welfare organisations take the position that exploiting and using animals is not in and of itself wrong – instead, it is the way that it's done that's bad. The issue is not, therefore, that we forcibly impregnate animals, or that we take newborn babies from their mothers and put them in solitary hutches. Or even that we slaughter animals, when doing so involves gas chambers and macerating them alive.

I highlight these examples, as all of these practices (and many more) are permitted by the RSPCA, the UK's leading animal welfare charity. To be fair, the RSPCA has stated that they would like gas chambers for pigs to be banned on welfare grounds, yet they also have guidelines for how to gas pigs in their pig welfare manual.[30] And the RSPCA still allows their welfare certification to be used when retailers sell flesh that has been taken from pigs killed in this way.

In other words, gas chambers are a welfare problem, but not enough of one that the UK's largest animal welfare charity will stop endorsing or making money from the practice. But don't worry, because if you go to the RSPCA Assured website you'll see the line 'RSPCA Assured is the only assurance provider dedicated solely to animal welfare.'

They also have guidelines for how to slaughter pregnant dairy cows, which include this chilling passage:

> If, for any reason, a foetus is found to be showing signs of life upon removal from the uterus (i.e. a foetus that has gasped and is now conscious), it must be immediately killed with an appropriate captive bolt or by a blow to the head with a suitable blunt instrument.[31]

Is this not the paradox of welfare accreditation? It states that animal wellbeing is its priority while allowing the fundamental immorality of exploiting animals to continue. If we think of a sweatshop producing fast-fashion clothes, the same welfarist logic in this situation would mean that the fundamental exploitation of the workers is acceptable as long as they are given food and water, a bed and a cushion to make sitting down more comfortable.

Do those welfare changes make the fundamental exploitation justified? Is depriving someone of their autonomy and their life ethical if we give them a cushion (or in the case of animals some straw)? Ironically, the reason why we call ethical fashion ethical is because it tackles the underlying issues that can be found in the garment industry. Yet we call meat ethical even though the fundamental issue still exists.

Now don't get me wrong, it's of course better not to farm an animal in a cage. But this line of argument presents a false dichotomy, as we don't need to farm them at all.

Some make the argument that people want to eat animal products and high-welfare is more achievable than abstinence, but since when did we think it was ethical to make compromises on the wellbeing of others simply because there are people who enjoy the consequence of harming them?

Despite all of this, you might be surprised to know that I'm not necessarily always against welfare regulations, as long as they are not the end goal. After all, even though I don't want animals to be farmed, I do want those who are to suffer less. I am, however, wary that some regulations could have negative consequences as well.

One of the downsides of welfare reform is the idea that it creates in the mind of the consumer. By making the problem the use of cages, for example, it implies that without them there is no issue. Free-range is a great example. People buy eggs that they believe to be ethical because they're not from caged chickens. However, the hens are still debeaked, still selectively bred, still often crowded together and still slaughtered at around 72 weeks old.

And even though the animal farming industries pay lip service to the issue of increased welfare, their actions tell a different story. A recent campaign to have farrowing crates banned for pigs in the UK was challenged and eventually stopped due to lobbying from the pig farming industry. Farrowing crates are cages so small that mother pigs can't even turn around in them. The National Pig Association (NPA) stated that such a ban would trigger a 'mass exodus' from pig farming[32] because of the extra financial costs that would be incurred.

This presents an obvious irony. If you ask the farmers opposed to this ban if the welfare of their animals is their number-one priority, they will say yes. No farmer says they care about profits over their animals. Yet the animal farming industries actively oppose legislation that is just trying to give animals the space to turn around and extend their limbs.

Ultimately, welfare does not make what we do to animals ethically justifiable, so buying animal products with a label on them is just an example of humane-washing – it makes us feel better, but the fundamental problems still exist.

'WE SHOULD ALL JUST RESPECT ONE ANOTHER'S VIEWS'

When you factor in the externalities that drive people's behaviours and rationalise their decision-making and beliefs, it is absolutely possible to have a respectful conversation with someone while challenging their views. However, 'We should all just respect one another's views' is often used as a way of saying, 'Please don't challenge my views at all.'

In fact, the notion of respect is the cornerstone of veganism. It's about respecting the desires, feelings and emotions of the animals who we currently show absolutely no respect to. A conversation I had with a student at Stanford University touched upon these ideas.

> **Jed:** I certainly respect your opinion that veganism is moral, and I'm not going to challenge it. I'm not going to say that you are wrong. I am just saying that we can co-exist.

> **Me:** Oh, we can co-exist. We do. The problem isn't you and me, because I'm not the victim and neither are you in this conversation. The problem is the animal. They can't co-exist in this scenario.

It's easy to say we should all respect one another when we're not the victims. One of the biggest challenges the animals face is that they are not able to properly represent themselves, so the debate around veganism becomes very human-focused. The lack of a victim right in front of us makes it easier to take the stance of 'live and let live' (or should that be 'live and let die' in this context?).

However, an oppressor saying to the individual they're victimising that they should respect their views clearly represents a completely imbalanced power dynamic. If we think of all the situations where someone is victimising someone else, is the oppression ever justified by the argument that we should all just respect each other's views?

Once we address the elephant in the room, or more aptly, the pig in the slaughterhouse, and recognise animals as the victims that they are, we can then factor them into this conversation and recognise that, even though they can't speak like we can, they are the most important interlocutor in this conversation.

The idea of respect is also used when people make arguments such as, 'I make sure to show respect to the animals I eat.' Or, as a student in Texas put it: 'I would like to see those of us who enjoy meat do it knowingly. And do it with respect. "I know I'm eating this steak. I respect the fact that this animal lived and died for me."'

The notion of respect implies mutuality. It involves factoring in the needs, desires, rights and wants of those we are engaging with. What we do to animals is the opposite of respectful. It's a one-sided and biased approach that basically says, 'I can do what I want with animals, and as long as I'm OK with it that's all that matters.' And to not be vegan means we are actively disregarding the wants, feelings and desires of animals, which is the antithesis of showing them respect.

THE ANTI-WOKE WARRIOR

'VEGANS ARE JUST MEMBERS OF THE "TOFU-EATING WOKERATI"'

Veganism has unfortunately been branded as 'woke', although nowadays it seems nigh on impossible for anything to avoid being branded as such. However, by using the word 'woke' to tarnish veganism, people are merely using thought-terminating cliché. If you say the reason you don't want to go vegan is because it's woke, that doesn't mean anything. It's simply a way for people to avoid engaging in any form of serious conversation about an issue and instead use identity politics as a reason to dismiss it.

Plus, outside of the fact that the term 'woke' has become a weaponised word that shuts down meaningful debate, there's no logic to what is branded in this reductive way. Is being against the slaughter of dogs at the Yulin dog meat festival also woke virtue-signalling? What about being against trophy hunting or eating animals like octopuses while they're still alive? And if these things aren't examples of woke virtue-signalling, why is opposing pigs being forced into gas chambers, or male chicks being macerated alive, or cows being anally fisted?

If veganism is fundamentally about reducing animal suffering and harm, and the word 'woke' is used to describe something as being bad (although for lots of people, being woke just means being alert to injustice and discrimination), we are fundamentally saying that it is bad to try and reduce animal suffering and harm. So is reducing animal suffering a fundamentally woke or bad concept that should be dismissed? If not, then why would trying to reduce animal suffering as much as we can be considered a bad thing? It's nothing short of outrageous that trying to reduce suffering, harm and sexual exploitation is branded as being negative.

This whole way of thinking is just so depressing. Humanity is supposed to value progress, yet in the time since historical figures such as Aristotle and Socrates displayed their oratory skills and ability to stimulate impactful and challenging conversations we have seemingly not evolved at all. Here we are, more than 2,000 years later, and our discourse and argumentation has regressed to the quality of playground children squabbling over whose lunchbox is the best.

So how about, instead of reductively labelling something so that we can justify why we don't like it, we actually take important ideas and have meaningful, intellectually sincere and genuine conversations about them so that we can work together to make the world a better place? Or I guess we could do what Piers Morgan does and eat a steak in front of a vegan, because who cares about societal progress when there's social media engagement to be gained?

'VEGANS ARE TOO FORCEFUL AND PREACHY'

One argument that is especially common is that vegans are too forceful. I've heard this so many times, although, to be fair, after the third block of force-fed tofu, I find people do tend to come around.

There are two dimensions to this argument: the first is that some people dislike certain forms of animal rights protests.

And, in all honesty, it's perfectly reasonable to find that some forms of protest don't resonate with you. You might see a clip of vegans demonstrating online and not personally find it effective in convincing you of anything. However, while you might not find the form of protest itself persuasive, that doesn't invalidate the cause behind it.

The second dimension is that you might meet a vegan who is abrasive and obnoxious. You might even go so far as to say that you simply don't like that person. However, again, finding a vegan annoying or just not liking them doesn't justify consuming animals as a consequence.

Advocates are basically salespeople. If we meet someone who is trying to sell us a product and they are really bad at their job, while it might impact how we view the product, it doesn't mean that the product is necessarily bad too. In a similar way, snake oil salespeople make a living selling products that are very bad but do so in a very effective way.

I've met many forceful and not very pleasant non-vegans over the years – does that in and of itself justify veganism? Would it be a good argument if I said, 'The reason why you should go vegan is because I once met a non-vegan who wasn't very nice to me'?

So although we might not always like the rhetoric or the manner in which it's is delivered, that doesn't exempt us from being vegan or engaging with the issues.

The truth is, people call vegans preachy simply for talking about it in any capacity. We are accused of forcing our views on people left, right and centre. I've even been accused of forcing my views on someone just because I responded to a question that they had asked me.

It is true that vegans can be very passionate and outspoken in their beliefs, and it is easy to dismiss the arguments by simply labelling them as being preachy or forceful. However, people should be passionate about protecting the planet and the lives of others. We should be passionate about challenging a system that is responsible for so much harm and suffering.

I concede that some vegans can be overly argumentative – primarily because of how much they care about the negative impacts of animal consumption – but nobody can force someone else to be vegan. Vegans aren't following non-vegans around replacing the meat from their lunches when they're not looking. We've not infiltrated every coffee shop and secretly replaced all the cow's milk with oat milk (although, if we had, we wouldn't say).

But the food we buy does dictate what others are forced to endure. When we buy animal products, we are forcing animals to live a life in which they are exploited, harmed, mutilated and slaughtered. A non-vegan can stop engaging with a vegan; a non-vegan can ignore a video of a vegan protest they don't like. Animals can't simply walk away from the situation they are in, and they can't ignore it either.

When we talk about being forceful, nothing is more forceful than controlling every aspect of someone's life. Nothing is more forceful than subjugating someone to your will. Nothing is more forceful than taking someone's life from them.

So when we compare veganism to non-veganism, while vegans might be passionate, maybe even judgemental, what is actually more forceful? Campaigning for veganism or exploiting sentient beings and ruining the planet in the process?

It's easy to turn a blind eye to the suffering that we are causing and then accuse anyone who attempts to encourage us to stop causing that suffering as being forceful – but we are the oppressors in this situation. If we were the victims, I'm sure we would feel very differently.

'EATING ANIMAL PRODUCTS IS MY PERSONAL CHOICE'

It is true that what we choose to eat is a personal choice. But having individual agency doesn't give us carte blanche to do whatever we want. If our actions were justified simply by us having chosen them, pretty much anything would be justified as a consequence.

A murderer chooses to murder, a rapist chooses to rape, a bullfighter chooses to kill a bull. Are these things morally justified as long as the person committing those actions made the personal choice to do so? Similarly, if someone hits someone else with their car, is it OK as long as they made the personal choice to do so? Logically speaking, this would make accidentally harming someone worse than choosing to harm them. Imagine this argument being used in court: 'Your Honour, I did indeed kill that man. However, in my defence, I would like to point out that I did make the personal choice to do so.'

Fundamentally, even the most heinous crimes are committed by those who choose to commit them. In fact, when someone who has done something particularly terrible was lucid and aware of what they were doing, this is viewed as making what they did even worse. I wonder why that logic doesn't apply when it comes to harming animals.

This argument in favour of personal choice also removes any notion of there being a victim from the scenario and makes the impact of our food choices comparable to the impact of what music we choose to listen to. We view them as being decisions that impact only ourselves. However, there are of course many victims in our food choices, and we should factor them into our decision-making.

'VEGANS WANT TO BRING IN LAWS AND PUT RESTRICTIONS ON PEOPLE'S FREEDOM'

We already have laws prohibiting us from eating certain animals. For example, prohibiting people from eating shark fins and dolphin meat restricts people's freedoms, as do laws in general. However, we have these laws because people having the freedom to cause harm to others is viewed as less important than protecting those who would be harmed.

This is the reason we have animal cruelty laws to begin with (even though there are some major oversights and contradictions in them) – because we recognise that non-humans also deserve protection. Many of the laws that vegans would want to bring in are not actually new laws at all – they are simply extensions of the ones we already have but applied consistently and logically to other animals who should also be protected in the same way.

The argument that vegans want to restrict people's freedoms was actually put to me by radio presenter Nick Ferrari when I was discussing veganism live on the BBC.

Nick: Might I ask you one question? If you had powers, would you seek to restrict people eating meat?

Me: No, I don't want to mandate things. I want to educate people to understand that there's a better, more ethical, more sustainable and ultimately healthier way that we can farm and use our land.

His question was an attempt to frame me as being an extremist who wanted to impose my beliefs on others. However, while it is true that I would like our laws to be consistent, the idea that my aim is to have laws imposed on society is naive at best and scaremongering at worst.

Outside of authoritarian regimes, laws tend to be reflective of societal attitudes. Politicians and political parties want to appeal to the electorate, so laws that are wildly unpopular do not serve to aid politicians in their desire to be re-elected and maintain legislative authority.

If our laws were to ever reach a point of consistency on the subject of animal cruelty, it would be in a society where the attitudes of the public reflected those laws. So the question of mandates and laws is really secondary to the question of advocacy, as it is through advocacy that public attitudes can change. In other words, as I said to Nick, education and swaying public opinion are the real means for creating societal progression.

Case in point, while we might have laws stopping people from kicking cats, the reason the majority of people don't kick cats is not because they're scared of the legal repercussions but because they believe it to be a morally repugnant thing to do. The law exists to punish those outliers who act in this horrible way, not to mandate and force a population of people who actually want to kick cats into submission.

Realistically, the same will be true if laws are brought in making farming animals illegal. Currently, such laws would lead to civil unrest, animosity and no doubt a severe rebuke at the next election, but things can and do change, and that's why I spend so much of my time trying to educate people about the benefits of veganism.

And none of this is to say that vegan advocacy doesn't push for any new policies; for example, many of us argue for a change in how we subsidise our food in order to help farmers transition out of animal farming (unfortunately, such initiatives have so far been ignored). However, these are not policies aimed at forcing people to stop eating meat but instead to address an imbalance in how public funds are currently being spent.

While the idea of vegans imposing laws on meat eaters is an effective way of trying to portray them in a bad light, in truth, meat eaters also believe in animal-cruelty laws. Vegans just want to reach a point where these laws are applied consistently.

'VEGANISM IS BEING PUSHED BY GLOBAL ELITES'

On a podcast in 2023, self-proclaimed 'freedom fighter' J.P. Sears, who uses the handle @AwakenWithJP, stated, 'Bill [Gates] and the World Economic Forum, like the globalist reptilian people, are so desperate to get people to stop eating meat. It just makes me believe more than ever that meat is important to eat, because I think they want to weaken people because weak people are easier to control.' Sadly, this kind of rhetoric is typical of the conspiracy theories that are levelled at veganism.

Later on in this book, I debunk the argument that plant protein is inferior and will make you weak (see 'The Amateur Nutritionist'), but the more extreme idea that meat will prevent you from becoming subjugated is also demonstrably and clearly untrue. I do, however, think it's important to note that criticism of the World Economic Forum (WEF) and the hoarding of wealth and indeed political power and influence by the ultra-rich is perfectly valid. However, criticism that is unsubstantiated or that contradicts reality actually deflects genuine concerns and undermines the discourse.

When you actually look more objectively at what is going on, there are some clear contradictions at play. For example, while it is true that Bill Gates has invested in plant-based meat companies, he has also invested in companies that are developing seaweed supplements for cattle,[1] and his foundation has given a grant to a company designing methane masks for cattle[2] (see 'The Misinformed Environmentalist' for my response to both of those methane inhibitors).

It seems strange that a man who wants to stop people from eating meat to make them weak would then also fund technology that would actually encourage people to continue consuming animal products. The reality is that Bill, like many ultra-wealthy individuals, is investing in the potential technology of the future and diversifying his portfolio for many different eventualities.

The UN is another frequent focus of these conspiracy theories. However, the UN's Food and Agriculture Organization advocates for animal farming and for further intensification of the system, and has actively tried to suppress reports on its negative impact.[3] At the same time, the UN's Intergovernmental Panel on Climate Change (IPCC) has been forced to water down published reports and remove recommendations for plant-based diets.[4]

This leads us to the question: who is really pushing what?

The reason the IPCC was forced to remove any recommendations for plant-based diets was because of lobbying and political interference from the meat industry and governments of large meat-producing countries. The truth is these conspiracy theories don't challenge the status quo or scrutinise the most politically and financially powerful industries or companies – in fact, they do the opposite. They might attempt to frame it that way by targeting high-profile individuals such as Bill Gates, but if his investment in cell-cultured or plant-based meat doesn't work out, he's still going to be ultra-wealthy, so he has very little at stake.

It is the animal farming conglomerates who benefit from these conspiracy theories, just as it is the fossil fuels companies who benefit from conspiracy theories about green energy. In other words, it is the most powerful, politically influential, morally bankrupt and harmful industries that benefit the most.

Problematically, a lot of the people behind the lobbying and political power of these industries are anonymous. We don't know who they are, what they look like, or even the organisations that they represent. They don't hold widely publicised and recorded forums where clips of them can be cut up and shared online. It's much harder to target or distrust something that we can't engage with. Conspiracies about Bill Gates and the WEF, on the other hand, are far simpler to push, because they involve highly identifiable individuals and organisations with logos and acronyms that are memorable and part of the public discourse.

While it might be tempting to gloss over these outlandish narratives, the issue here is not that people like J.P. Sears are on the precipice of discovering some deep conspiracy, but that their views can trickle down into mainstream thinking and then be viewed as legitimate. For example, the controversial psychologist Jordan Peterson stated in a video attacking plant-based diets that the future would be 'plants and bugs for you and your children, and the sooner you get used to that, or else, the better'.

A UK farming organisation called No Farmers, No Food has also amplified these ideas, resharing on X that 'Farmers stand between us and WEF's desire for us to EAT BUGS, own nothing and be happy.'[5] And James Melville, who started the campaign group, has reportedly reshared statements such as 'Farmers across Europe are mass protesting the globalists trying to crush them.' Even the name No Farmers, No Food implies there is some concerted effort to eliminate farmers. But nobody wants to get rid of all farmers, so even the name is hyperbolic and a means of manufacturing fear around something that isn't real. And yet, at a protest in 2024, then

Prime Minister Rishi Sunak stood alongside this campaign group to protest against the Welsh Labour government's plans to incentivise Welsh farmers to put aside just 10 per cent of their land for things like woodland and wildlife.[6]

Similar problems have emerged in Europe, with certain sections of the farming community and wider society making farmer protests a symbol of resistance against a global elite. The truth is these narratives are a very cynical ploy to use fear to push a political narrative and the financial interests of one of the most powerful and wealthy industries on the planet. It also positions farmers as victims and on the frontline of an existential battle that pertains to the very freedom of humanity. The scary thing is, though, that this kind of messaging does resonate with a lot of people. A report looking at the spread of misinformation around plant-based diets online showed that the biggest category of misinformation was about conspiracies involving elites and a 'Great Reset'.[7]

By tying in plant-based diets with theories about subjugation and the removal of autonomy, people are being scared away from veganism. Not only does this prevent meaningful and necessary change, but it also alienates and isolates many farmers whose genuine concerns are being pushed to the side and ignored in favour of extreme ideologies. The truth is these views are not shared by all farmers, and the farming community is being exploited as part of a larger ideological drive to further extremist and often far-right views that are being put forward by opportunists who are not involved with farming and who are more interested in garnering attention for themselves than they are with actually helping farmers.

When the IPCC or the World Health Organization warn of the dangers around animal products, they are drawing these conclusions from a huge body of research and evidence, conducted by organisations and scientists from all over the world. This isn't some opaque behind-closed-doors scheming.

These theories invert reality and amplify pseudoscience and propaganda, while tarnishing huge studies that involve experts from all around the world. And the thing is, the scientific literature regarding the healthfulness of plant-based diets and the associated risks of animal products is publicly available, as is the environmental science. If you have access to the internet, you have access to the evidence. But sadly, and not necessarily through any fault of their own, people are not scientifically or media literate, and eye-grabbing headlines or social-media memes and posts that play to our fears and create doubt are more persuasive and easily digestible.

By making things like government legislation, mainstream acceptance and scientific evidence the centre of their distrust, the more that the merits of a plant-based diet become recognised, the more the views of conspiracy theorists are reaffirmed. Evidence for their distrust will be found any time a government makes their nutrition guidelines more plant-based, or any time a landmark piece of research promoting a plant-based diet is released, or any time a piece of animal farming legislation is discussed.

So no, there isn't some great conspiracy to use plant-based diets to subjugate people. There is, however, a very powerful, well-organised and well-funded industry that is working to censor good science, disseminate pseudoscience, shut down political discourse, and ensure its vested interests are protected.

'EATING ANIMAL PRODUCTS IS A SYMBOL OF POWER AND STATUS'

One of the most notable times I heard this argument was when I was at Texas A&M debating veganism with a guy called Vladimir. The conversation took a slightly unusual turn when I asked him what would happen to our species if we stopped eating animals.

> **Vladimir:** I mean no disrespect. I think we would end up softer, weaker, and other nations that still consumed meat would overtake the United States because they can harness something very primal and very dangerous that we have lost the ability to harness.

> **Me:** So, just to summarise, if Americans were vegans, they would be too weak to stop a communist invasion?

> **Vladimir:** In a sense, yes. I do believe that we need a society that eats meat so that we can still harness some of the primal rage and power to protect ourselves.

To be fair to Vladimir, I was the one who mentioned a communist invasion, but it was fairly obvious which nations he was worried about overtaking the USA. Plus, if you were asked to think of one place in the world where the people were likely to be afraid of both veganism and communism, Texas would probably spring to mind.

While Vladimir's claims are outlandish, we do often view the killing and consumption of animals as a symbol of being strong and as a means through which we can display our dominance and affluence. On the flip side, we then view the eating of plants only as symbolising weakness and a lack of prosperity. A lot of this comes down to how meat-eating has existed historically, with the wealthiest and most powerful people being the ones who could consume animals in abundance. During the twentieth century, this idea was reinforced as the increased consumption of animal products became a symbol of post-war prosperity and growth. This has led to people viewing the consumption of meat as a way to elevate their perceived status within society.[8]

The perception of meat as being the ultimate source of protein (see 'The Amateur Nutritionist') and also as being a sign of masculinity further plays into the idea that the consumption of meat allows us to harness some raw and primal power. Of course, this is total nonsense and is purely a societally constructed idea based on historical power dynamics, advertising and false connotations. You can be physically strong and not eat animal products just as you can be physically weak if you do eat them.

Some people might argue that it's not actually physical strength that's being alluded to here, rather emotional or psychological strength. However, if you have to pay for animals to be killed in order to feel powerful or emotionally satisfied, it's not meat that you are missing. Plus, vegans don't just roll over and allow everyone to walk over them. You can't have it both ways: either we're forceful, militant and want to impose our beliefs on others, or we're weak and soft and unable to defend ourselves because we don't eat animals.

Using the subjugation of animals as a means by which we can elevate our own perceived status is not only tyrannical but also misguided. Eating animal products doesn't change who we are fundamentally as individuals. In fact, the research shows that eating meat to gain status is done precisely because there is a perceived lack of social status in the first place.[9] In other words, eating animals to make oneself seem more powerful is paradoxical, as those who are comfortable with their social status don't feel the need to eat animal products to substitute something they perceive to be lacking.

THE WELL-INTENTIONED LEFTIST

'EATING ANIMAL PRODUCTS IS AN IMPORTANT PART OF CERTAIN TRADITIONS'

It is true that many traditions include the consumption of animal products, but similar to the culture argument in 'The Social Conformer', just because something is a tradition does not make it morally justified. Examples of traditions from around the world that few people would judge to be morally defensible include honour killings, female genital mutilation and the persecution of homosexuals.

However, rather than just claiming that something is moral because it's traditional, one of the main reasons people appeal to tradition in defence of eating animal products stems from the fear that trying to prohibit certain traditions is another form of Western colonialism. But if you follow this logic, you could argue that it is colonial to demand that the people in Pakistan who carry out honour killings should subscribe to our beliefs that they are wrong. Or that it is a form of colonialism to try and prohibit the African cultures that perform female genital mutilation from doing so. But these practices are wrong and should be stopped even though they mainly occur outside of the Western world and are considered by some to be important traditions.

There are of course traditions that should be defended – peaceful worship, for example – but the line is drawn when such practices cause suffering to others and involve victims. If a dance, for instance, involved the brutalisation of someone, it would be unethical regardless of its cultural significance.

We can think about issues of animal suffering in traditions in a similar way. The celebration of religious festivals and other culturally significant events should be protected, but not if they involve the brutalisation of an animal. The argument is not that cultures shouldn't be able to celebrate their identities through their traditions, just that they shouldn't come at the cost of someone else.

While many traditional and cultural practices do involve animals in some way, the meaning behind them is usually far greater than the use of the animals. Although we associate Christmas with the consumption of turkeys, for example, eating turkey flesh is not what Christmas is really about. Removing a dead turkey from the dinner table does not erase Christmas, as it is actually about celebrating the birth of Jesus or, from a secular point of view, about family, communion and giving to others. None of this is lost if we stop killing animals to mark the occasion. Similarly, vegan Muslims are still Muslims even if they don't sacrifice a goat for Eid. Because Eid is about offering a sacrifice to show gratitude to Allah, some vegan Muslims instead donate money to help people less fortunate than themselves. In this way, the essence of the tradition is maintained, which is surely the most important thing.

'VEGANISM IS RACIST'

Rather impressively, veganism is able to be both an example of the woke, snowflake PC culture gone mad and also racist, ableist, anti-indigenous and colonialist all at the same time. In other words, it manages to draw ire and condemnation from both the left and the right simultaneously. So let's dig a little deeper into one of the main leftist arguments against veganism, which is that it is racist. As one blogger put it, 'It is outright racist. In fact, at this point, being a white vegan is practically intrinsically racist.'[1]

There seem to be two main versions of this argument. The first is that veganism is racist because people of colour are disproportionally more likely to live in poverty. In fact, in the UK, it's estimated that Black and minority ethnic people are twice as likely to live in poverty than white people.[2] The argument continues that it's therefore unfair to expect people of colour to be vegan because it is more expensive (for more on whether this is actually the case, see 'The Practicalist').

The second reason why some people claim veganism is racist is because of the historical struggles that people of colour have faced (and continue to face). As a result, it is argued, they don't have the emotional bandwidth to be vegan, and until we live in a society free from racism, advocating for Black people to care about animal rights is therefore unreasonable. However, I've also seen it expressed by some Black vegans that this line of thinking minimises the racist experiences that they've had, as they still have the emotional bandwidth to be vegan. The experience of Black people is not monolithic and arguments that suggest Black people as a demographic can't or shouldn't do something appears reductive and neglects the autonomous nature of people, regardless of their ethnicity or skin colour. And the societal challenges that people of colour face are not diminished in their severity by the acknowledgement that what we do to animals is a moral and ethical issue in its own right.

It's very important to understand the social barriers that can exist to veganism and why some people of colour have reservations about it. However, this argument is often perpetuated by white people who are looking to advocate on the behalf of people of colour or explain their situation for them. There is something particularly uncomfortable about white people stating that people of colour shouldn't go vegan, as not only does their perpetuation of the idea that veganism is racist carry the potential to alienate people of colour who might be interested in it, but it also erases their voices from the debate. White people stating that veganism is racist to vegans of colour is clearly problematic.

Veganism is an attempt to reduce animal suffering, exploitation and death. It is a movement that wants to grant non-human animals relevant and applicable rights. Nothing about the aspiration of treating animals with more kindness, dignity and respect is intrinsically anti-Black or racist.

And actually, polling in the USA shows that Black people are more likely to give up meat and also to become vegan. It's estimated that while 3 per cent of people in the USA are vegan,[3] 8 per cent of Black people are, with Black people making up the fastest-growing vegan demographic in the country.[4]

There are several reasons this is believed to be the case, including the promotion of veganism by culturally important Black artists, a growing awareness of the link between animal products and chronic diseases that disproportionately impact Black people, and growing anger about the existence of food injustice, as well as an increasing awareness around the impact of animal farming on the animals and the environment.

When white people who are not impacted by many of these factors in the same way then step in and say that veganism is racist or that it's for white people only, they are invalidating the experiences that Black people have as a result of our current food system, and they're impacting the progression of something that positively benefits communities of colour as well.

Angela Davis, the civil rights and political activist, philosopher, and academic said this about veganism:

I usually don't mention that I'm vegan. But that has evolved and I think it's the right moment to talk about it because it is, I think, a part of a revolutionary perspective. How can we not only discover more compassionate relations with human beings but how can we develop compassionate relations with the other creatures with whom we share this planet? And that would mean challenging the whole capitalist industrial form of food production.

Veganism isn't a barrier to the eradication of injustice and oppression, it's an extension of it.

'VEGANISM IS A FORM OF CULTURAL APPROPRIATION'

This argument is based on the fact that there have been cultures in history that have abstained from eating meat, although many of them consumed dairy, and also that many vegan staples are culturally significant foods in Asian and African countries.

This is another example of vegans getting targeted with two contradictory positions. Veganism is criticised because there's no historical precedence for it (as discussed in 'The Historian') and then, as this example shows, it's also criticised for ignoring the historical roots of abstaining from animal products.

A plant-based diet certainly includes a variety of foods from all around the world, and some plant-based staples are from non-Western cultures. But why is this a problem? Are people only supposed to eat the foods from their culture? Vegans don't think they own the intellectual rights to a lentil curry or a tofu stir-fry. They enjoy eating them because they're made from plants and are delicious. This is really a straw man argument – it's criticising vegans for something that they don't believe. Vegans aren't saying that there was nobody before the twentieth century who abstained from eating animal products. And vegans know a lot of the food they eat is from all around the world.

The problem with the arguments used to support the accusations that veganism is racist and a form of cultural appropriation is that they contradict one another. On the one hand we're told that veganism is racist because it's a white person thing and that people of colour shouldn't be expected to care about animals or the issues around animal consumption. Yet, on the other hand, we're told that veganism is a form of cultural white-washing because it centres on white people and ignores how abstinence from certain animal products has been a part of non-white cultures for thousands of years. This contradiction only further emphasises the weakness of both positions.

'VEGANISM IS ANTI-INDIGENOUS AND COLONIALIST'

The question around veganism and indigenous people really comes down to those who can be vegan and those who can't, as there are many indigenous communities around the world who are undeniably unable to be vegan. For them, my response is very straightforward: if you are unable to be vegan, you are unable to be vegan.

However, what about indigenous people who have access to plant-foods or who can shop in supermarkets and eat at restaurants, and therefore who can be vegan? Well, the same logic applies to them as it does to any other community or group of people. Being indigenous doesn't give someone the right to cause harm to animals when they can avoid it. From the animal's perspective, it makes no difference if they are killed for an indigenous person or a non-indigenous person.

Does that mean that vegans shouldn't be considerate about how indigenous communities have been impacted by colonisation and had aspects of their cultures denied to them while also being forcibly assimilated into other cultures and societies? Of course not. But do past transgressions inflicted on indigenous communities mean that animals should have to die? I would argue not, especially as animals are not responsible for those transgressions. Why should they suffer for crimes they didn't commit?

The idea that indigenous people should be exempt from this conversation plays into the idea of the soft bigotry of low expectations – in other words, that we shouldn't expect members of disadvantaged groups or minorities to meet the same standards as other people. But why shouldn't indigenous people also engage in these conversations, and why shouldn't they also have to address the same moral concerns that there are around eating animals? After all, vegans are not forcing anyone to be vegan – we're not violently coercing indigenous people into being vegan against their will. If an indigenous person decides to kill an animal or eat animal products, they're making that choice of their own volition.

This argument also erases indigenous vegans who still celebrate their identities and cultures but do so while trying to minimise the harm that they cause to animals at the same time. Are they betraying their indigenous heritage? Are they any less indigenous because they're vegan? If not, then that means an indigenous person's connection to their culture is not inextricably tied to the killing of animals.

Indigenous cultures are defined by so many other things, including language, music, dance and spirituality. None of these things are compromised by allowing an animal who does not need or want to die to live. And importantly, it is these other, core aspects of culture that should be defended, respected and honoured (just like those aspects of other traditions discussed earlier in this chapter).

The reality is, the people who most often seem to bring up indigenous culture as a defence for animal consumption are white people who are not vegan themselves. But appropriating other people's culture to justify our own habits is not respectful. It's an attempt to hide our behaviours from moral scrutiny by aligning ourselves with a community and a group that we are not part of.

We should also consider that the leading cause of deforestation in South America is animal farming, and there are direct links between the animal products produced there and the displacement of indigenous communities in those countries. These products are then exported to places like the USA, the UK and the EU (more on this in the response about soya farming in 'The Misinformed Environmentalist'). So, if anything, the most valid arguments in relation to indigenous people are those that are actually in favour of veganism due to the very real impact that animal consumption in the West has on the lives of indigenous people in South America in particular. Not to mention the impact that the climate crisis has on indigenous communities all around the world, further demonstrating that veganism can directly benefit indigenous people.

'VEGANISM IS ABLEIST'

Veganism as an ethical philosophy is not ableist (the word used to describe discrimination against disabled people). It is simply the pursuit of granting non-human animals applicable and relevant rights.

The ableist argument generally focuses on practical issues,

not philosophical ones, and the real issue is not with veganism but current food accessibility and food policy. However, the practical issues that some disabled people face are then extrapolated to the point that some people claim veganism to be ableist, when in reality the point is that there are some disabilities that create practical barriers to veganism for some people.

Now, can a vegan be ableist? Yes, of course they can. Making claims that someone wouldn't have their disability if they didn't eat animal products or that they shouldn't be taking medication because it's been tested on animals is ignorant, arrogant and ableist. But that doesn't mean veganism is intrinsically ableist.

Making sweeping claims about veganism being ableist erases disabled vegans and minimises what they are doing. And not only is it invalidating to vegans who are disabled, it's patronising to disabled people who could be vegan but are instead being told that because they have a disability they should be sheltered from the discussion. Disabled people who are perfectly capable of critically reflecting on their actions and choices should be given the space to do so.

There's also a danger of grouping all disabled people together to the point that they become a generalised abstraction. There are many types of disabilities that affect people in lots of different ways. Saying veganism is ableist removes the individuality and personalised experiences of each disabled person.

What is ableist is classifying disabled people as existing outside of the general population, or of being unable to engage in ideas and discussions that able-bodied people can. And even if someone's disability means they are unable to go vegan, that doesn't mean that they don't agree with the merits of veganism or that they would want people who can be vegan to not be vegan.

I've also spoken with disabled vegans who have made the point to me that the language associated with ableism often overlaps with the language that humans use to demean animals. In particular, a disabled vegan told me that one of the things that led them to veganism was recognising that a lot of the rhetoric they were using to justify not being vegan was actually similar to the disparaging things that other people had said about them.

So instead of using the situations of others as a reason to continue eating animal products, let's be intellectually honest and recognise that while some people have disabilities that create challenges for them to be vegan, this doesn't diminish the importance of veganism as an ethical pursuit for those of us who can.

'VEGANISM IS CLASSIST AND PRIVILEGED'

Calling veganism privileged always strikes me as peculiar. After all, privilege is in many ways relative. Compared to some people, having a home to live in and a roof over your head is privileged. Having access to clean water is also a privilege, as is having access to an education. So, if you're able to read this book, you have a form of privilege that many people around the world don't. But that doesn't make reading any less important or something that we shouldn't strive for all people to be able to do.

The same is true of veganism. Of course, having autonomy over your food choices is a privilege. Of course, being in a position where you have different food options to pick from is a privilege. But none of that invalidates the merits of veganism.

I had a student at Stanford University defiantly tell me that I needed to check my privilege because I was having conversations with the students on campus about veganism. It feels almost ridiculous to have to write this, but if you study at Stanford University, you don't get to play the privilege card.

But is veganism classist?

The pervasive myth that vegans have to be from a certain socioeconomic background is also misguided and unsubstantiated. This argument just completely erases working-class vegans and all the other vegans who don't tick the box of being wealthy or middle-class. Why would we segregate working-class people from the conversation? Why shouldn't they be factored into this discussion and given the opportunity to determine the morality of their actions?

Furthermore, just because expensive health and wellness products can be vegan and are found in places frequented by wealthier individuals does not mean that veganism is classist. This would be akin to claiming that eating meat is classist because there are steakhouses that only wealthy people can afford to eat at. And while the financial argument against veganism is addressed in more detail in 'The Practicalist', research has shown that a whole-foods plant-based diet in high-income nations can actually cut a third off your food budget.[5]

In areas where there is a genuine lack of plant-foods, meaning that someone can't be vegan, that is a failure on a societal and food policy level to ensure people have access to food. This is not the fault of veganism or an example of veganism being classist. This is especially true when you consider that the food and social policies we currently have were not designed by vegans or designed with the intention of ensuring that everyone has access to healthy, affordable plant-foods.

It's not vegans who created food deserts or vegans who created the subsidy and land-ownership policies that perpetuate the current system of animal farming. It was the legislators and landowners who benefitted and who continue to benefit from this system who created it and continue to prop it up.

Vegans don't blame people from lower socioeconomic backgrounds for the hardships they face. Instead, they wish to create a fairer society so that everyone has access to plant-foods, thus equalising food accessibility. In this regard, veganism is about challenging the dominant paradigm that has been created and perpetuated by the ruling classes. It's about recognising that if we want to create a fairer food system, veganism is a huge step in the right direction to achieving that.

Plus, the only way that veganism will become more accessible, thereby allowing more people to make the change, is if those of us who can become vegan do so and further normalise it and make it more accessible and affordable through simple supply and demand.

Too much anti-vegan rhetoric is centred around people attempting to appropriate situations that others are in so that they can tarnish veganism. It's an attempt to justify why they're not vegan but hidden under the facade of defending others. But when vegans say that people should go vegan, they're not talking about people from coastal regions of West Africa who are reliant on fish. They're not talking about homeless people reliant on food banks and soup kitchens. They're talking about people who can be vegan but choose not to be.

'THERE'S NO SUCH THING AS ETHICAL CONSUMERISM'

The argument against veganism that there is no ethical consumption under capitalism is simply an appeal to futility. You only have one petrol-guzzling pick-up truck?

Well, then, you should get more. You like buying clothes? Well, you're in luck, because you can buy as much fast fashion as you want. Wait a minute, you turn your electrical devices off when they're not in use? You can stop doing that, and while you're at it, plug in some more devices at the same time. After all, there's no ethical consumption under capitalism so just do whatever you like and cause as much harm as you can as you'll never be perfect.

Astonishingly, one left-leaning YouTuber stated that 'owning child pornography and buying a T-shirt are equally immoral actions'. If this doesn't sound right to you, then you already see how flawed the logic of this argument is. After all, even in an imperfect system there are degrees of harm. So we should still make decisions that reduce the harm that our actions cause.

No economic system is perfect, and simply by being alive we will all have some form of impact on the planet. It seems likely, therefore, that the concept of 'no ethical consumption' will always be true to some degree or another. So by all means critique social and economic policies, but let's leave the animals out of it.

'SHOULDN'T WE TRY AND END ALL HUMAN SUFFERING BEFORE WORRYING ABOUT ANIMALS?'

A student who went by the name of Q in Dallas asked me if I had considered looking at things from the other direction and focusing on humans first. My answer was simple: are we not able to do both at the same time?

Going vegan doesn't mean we can't tackle human suffering too, or even that doing so will be slowed down if we tackle animal suffering as well. If we are campaigners for issues related to human suffering, we can still dedicate our time to the issues we are working on and simply be vegan at the same time.

This argument is what is called a fallacy of relative privation, meaning that we make an appeal to an issue that we believe is worse; for example, 'What we do to animals is not *as bad* as what's happening to people.' (This is closely linked to the concept of 'what-aboutery', which is discussed in more detail in 'The "What-Abouter"'.) Even if you don't think that what we do to animals is the most pressing issue our species currently faces, that doesn't mean it's not an issue that needs to be addressed.

We could say, 'Shouldn't we do something else?' in pretty much any situation. If you campaign to end homelessness, someone could say, 'But shouldn't we solve global famine first?' Or if you campaign for better funding for schools, someone might say, 'Yeah, but shouldn't we end all wars before we start thinking about school funding?' It's a slippery slope when we start ranking suffering and injustices. Unfortunately, there are many problems around the world that need to be addressed; however, we can work to tackle multiple problems at the same time.

It's also worth mentioning that there are many issues – such as war or famine, for example – that require us to have some form of active involvement, whether that be something like marching or protesting, in order to effect change. We can't make a lifestyle swap and end war. On the other hand, what we do to animals is an issue that can be tackled passively. We don't have to become campaigners and activists to make a difference – we just need to change how we eat and consume.

In addition, it's not as if veganism only tackles the issues around what we do to animals. The animal-exploitation industries also contribute to the suffering of humans. For example, slaughterhouse work is well documented as being one of the most dangerous jobs that someone can do.[6] Not to mention that it has been linked to a whole host of mental health issues, such as PTSD, alcoholism, drug abuse and suicidal thoughts.

Slaughterhouse work doesn't just affect the workers. A review of crime and arrest statistics in 581 US counties showed that the larger the slaughterhouse, the higher the number of arrests and reports in that area, with the largest slaughterhouses being associated with a little under two and half times the number of arrests when compared to areas without slaughterhouses. Particularly noteworthy were the increased rates in rape arrests and sex offences.

The study controlled for variables and also compared slaughterhouse communities with those situated around other industries, such as iron-and-steel forging and truck-trailer manufacturing, but not only were they not associated with a rise in crime, in some cases they were even correlated with a reduction in the crime rates. This suggests that an increase in violence can't be blamed on factory work itself, or on the people most likely to take such jobs, but instead on slaughterhouse work specifically.[7]

There are also issues in the seafood industry, with tens of thousands of workers being victims of debt bondage, forced labour and human trafficking.[8] We might think this has nothing to do with us, but products from this exploitation have been traced to supermarkets and food establishments in Europe and the USA.[9]

Factory farms can also be harmful to the workers, as well as for those who live in the surrounding areas. Workers are around cesspools and faeces all day and are exposed to bacteria, harmful gasses and potential viruses like bird flu. The air pollution from farms also impacts local communities around them, and it's estimated that nearly 13,000 people die a year in the USA alone because of their proximity to animal farms.[10]

These are just some of the examples of how humans could be helped by shifting to a plant-based food system.

A meat eater might counter by pointing out the exploitation of humans involved in plant-based agriculture. After all, it's not as if plant farming is devoid of human exploitation. But that's not an argument against veganism, it's an argument in favour of an overhaul of our agricultural system in general and creating a plant-based food system that is better for humans and non-human animals alike.

Just like before, we fall into the trap of an appeal to futility by holding veganism to a standard of perfection and then using the fact that it's not perfect to continue perpetuating a system that is much worse. This argument is self-defeating, as not being vegan doesn't solve the problem of human exploitation in arable farming, especially as human exploitation exists in arable farming for animal-feed production too. Am I supposed to stop being vegan and eat both plants and animals, thus contributing to more human *and* animal harm? What exactly will that solve?

Plus, considering animal farming is one of the leading drivers of the environmental crisis and a major contributor to things like deforestation and freshwater scarcity, eating a plant-based diet has a beneficial impact on the lives of humans impacted by those problems as well. Just because there are certain plant-foods – cocoa, for example – that have human rights implications attached to them isn't a reason to not be vegan. You can buy ethical chocolate instead.

Our current plant-farming system wasn't created with the intention of making it as ethical or as sustainable as possible. So holding vegans accountable for the failings found within plant-farming that they didn't create is not only unfair, it stops us from making the changes necessary if we are ever to have an ethical and equitable food system.

THE MISINFORMED
ENVIRONMENTALIST

'ALMOND MILK IS BAD FOR THE ENVIRONMENT'

OK, let's talk about almonds. Because, let's face it, we kind of have to.

Almonds get a lot of criticism, especially from certain sections of society.

> **Piers Morgan:** Do you drink almond milk? . . . I think you're a hypocrite, like a lot of vegans I've met.

The main criticism of almonds is their water use. (There are also concerns about their impact on bees, but that particular argument is addressed in 'The "What-Abouter"'.) So let's tackle water.

Eighty per cent of the world's almonds are grown in California. Because California is a drought-riddled state, the assumption is that consuming milk from almonds produced there is bad because of the freshwater used. And there is a kernel of truth to this, because it's estimated that tree nuts in California, which include almonds, pistachios, walnuts and others, use around 16 per cent of the state's total agricultural water supply each year,[1] with almonds making up around 9 per cent on their own.[2]

However, another 16 per cent of California's agricultural

water footprint is just for alfalfa, which is used as feed for cattle[3], with 70 per cent of that being used as feed for dairy cows.[4] So just one feed source for dairy cows uses more than 11 per cent of California's agricultural water footprint. California produces about 20 per cent of the milk supply in the USA.[5] So just the alfalfa that's used as feed for dairy cows in California, which in turn produces just 20 per cent of the US milk supply, uses more water in California than the almond industry, which supplies 80 per cent of the whole world's almond demand.

A study looking at freshwater use in the Western US found that beef and dairy consumption is the 'leading driver of water shortages in the region'.[6] And a different study that looked at the impact of plant milk compared to dairy found that, on average, cows' milk uses 70 per cent more water than almond milk, is responsible for three and a half times more emissions and takes up nearly 17 times more land.[7] Basically, if you live in North and South America, cows' milk is much worse than almond milk from an environmental perspective. I also think it's exceptionally ironic that vegans get blamed for all the problems of almond production in California, yet nobody calls out non-vegans for consuming almonds.

Now does that mean that almond farming is fine just because there is something worse than it in California? No. Tree nuts do still require a lot of water, especially when compared to other plant-foods, and this is still an important consideration that should factor into our decision-making. Which brings me to the next point.

Some people argue that comparing global averages isn't fair to those who are producing cows' milk with less environmental impact. This is the case that dairy farmers in the UK make. Let's look at this from an emissions perspective. Dairy produced in Europe is responsible for about 30 per cent fewer emissions than the global average.[8] However, that's still about three times higher than the global average of emissions produced for the same quantity of almond milk.

It is also worth noting that nuts can actually have a positive impact on greenhouse gas emissions. This is because they grow on trees that sequester carbon dioxide, and this process, coupled with nut plantations being grown on certain areas of land, can lead to emissions savings due to what is called positive land use change.[9]

Obviously the UK gets a lot more rain than California, so the UK agricultural system is able to use that to reduce the amount of surface and groundwater that it needs. Irrigation does however occur in the UK dairy industry, as well as to grow things like cereals and grass that are used for animal feed.

It's reported that the dairy sector uses around 20 per cent of the UK's freshwater resource, which is the highest percentage of any industry in the food and drink sector.[10] And while the National Farmers' Union (NFU) doesn't want British farming to be unfairly demonised, it also doesn't like to admit that water consumption data from 53 dairy companies across the UK revealed that 60 per cent of sites were using water inefficiently.[11]

Furthermore, with the high quantities of feed imported from abroad, including feed used for dairy cows, focusing on just the UK water footprint of animal products is not truly reflective of British animal agriculture usage. Research has shown that around 75 per cent of the UK's water footprint for food and feed is located abroad.[12]

But even if the amount of freshwater used to produce dairy in the UK is less than that used in California, the same is also true for almonds. All the negative press around almonds comes from their production in California. But we don't have to buy almonds or almond milk from California. Almonds grown in Europe are done so using more traditional rain-fed conditions, meaning the freshwater use and irrigation of almonds in Europe is lower than in California. In the UK you can get almond milk from almonds grown in Europe by supporting brands like Alpro, Provamel and Plenish.

On the one hand dairy farmers don't want us to group all dairy farming together, but on the other hand they group all almond farming together and accuse it of being the same. And it doesn't end there. The biggest driver of river pollution in the UK is animal farming. And which sector is responsible more specifically? Yes, you guessed it, the dairy industry is the biggest culprit, having been linked to half of all farm-related river pollution in the UK.[13] In Devon, the biggest dairy-producing county in the UK, 9 out of 10 dairy farms failed to comply with regulations, with two-thirds of them actively polluting rivers.[14]

But don't worry because the NFU has a rigorous and foolproof plan to deal with the problem. They want the industry to take 'voluntary action through industry-led initiatives'.[15] Because nothing is more effective than an industry regulating itself through action that the industry members can then decide if they want to voluntarily enact or not.

And just a point on water usage more broadly, research has shown that in European countries, including the UK, switching to a meat-free diet could reduce the water footprint of our diets by as much as 55 per cent.[16]

However, even if almond milk were the most environmentally destructive milk you could purchase, the solution wouldn't be to pay for animals to have their reproductive systems and mammary glands exploited. The solution would be to just drink one of the dozen or so other plant milks available.

But wait, isn't soya farming destroying the rainforests in South America? I'm so glad you asked . . .

'SOYA FARMING IS BAD FOR THE ENVIRONMENT'

Poor soya, the most misunderstood of all crops, blamed for everything from man-boobs and infertility (see 'The Wishful Thinker') to the destruction of the Amazon (more on that in a moment). A humble, protein-rich legume, it must look at the significantly less derided lentils with envy.

Although, it does make you think. Why would one of the best sources of plant-based protein and calcium, which is loaded with other nutrients and is used to make vegan staples like tofu and tempeh, but can also be used to make plant-based milk, cheese, yoghurt, cream and meat replacements, be so derided and attacked? It's almost as if it poses a direct threat to something.

Anyway, conjecture aside, is there any truth to the criticism about soya? The basis for these criticisms stems from the deforestation that is linked to soya farming, and in truth, soya farming is a leading driver of deforestation in South America. So case closed, right? Well, not exactly.

More than three-quarters of the soya produced globally is used as animal feed, the majority of the rest is used in biofuels, oils and other industrial products and just 7 per cent of whole soybeans are consumed directly by humans.[17] It is this 7 per cent that is used to make soya milk, tofu, etc., although it's worth pointing out that non-vegans consume foods made from whole soybeans as well, such as soy sauce, miso and edamame. Human food therefore makes up a small percentage of what soya is used for and an even smaller percentage is used to make so-called vegan plant-based products.

When we turn our attention to the soya produced in Brazil, 95 per cent of it is used for animal feed.[18] In other words, not only is the vast majority of soya grown globally used as animal feed, but when looking specifically at the area of the world where vegans get criticised for contributing to deforestation, the percentage used as animal feed is even higher still. Not to mention that the industry that is actually the leading driver of deforestation in Brazil is not even the soya industry, it's the cattle industry.[19] Products from the Brazilian cattle industry are sold all around the world in the form of beef and leather. A report into the cattle industry found that companies including Nike, Adidas, Dr Martens, H&M, New Balance, The North Face, Timberland, and the list just keeps going, have supply-chain links to the Brazilian cattle industry.[20]

But what about soya? Consumers might not think they're supporting the production of soya in South America – after all, if you live in the UK and buy British pork, chicken and dairy, you probably think you are supporting British farmers. But just because an animal was farmed and slaughtered in the UK doesn't mean their feed only came from the UK. At least 90 per cent of the soya imported into the UK is used as animal feed.[21] In total the UK imports somewhere around 3.3 million tonnes of soya a year, which requires an area of land equivalent in size to 11 Greater Londons. Of this number, 77 per cent comes from locations with high deforestation risk.[22] This is the soya being used as animal feed.

The two biggest users of imported soya are the poultry and pig industries. However, even low estimates show that the dairy industry uses 8–10 per cent of the UK's soya consumption.[23] So buying dairy over soya because we're worried about deforestation is actually having the opposite impact to what we think.

Soya is also grown in countries across Europe. It is this soya that is most often used by plant-based companies, something you can more often than not check on their websites. You can't do that with the cows' milk, eggs and animal flesh we buy, and it is these products that are the true culprits when it comes to deforestation.

'VEGANS EAT PALM OIL'

Palm oil is a notorious food because of the deforestation and loss of wildlife that is associated with it, so let's dive a little deeper.

Palm oil use has seen a huge increase in the past 50 years or so, with production being 35 times higher in 2018 compared to 1970.[24] The reason for this meteoric rise is due to the efficiency of palm oil compared to other vegetable oils (more on this in a moment). However, even though it is the most efficient oil crop, the increasing demand for it has caused significant problems, with the land used to produce it having more than quadrupled since between 1980 and 2018,[25] with 84 per cent of the world's palm oil coming from Indonesia and Malaysia.[26]

Globally, 29 million hectares of land is used to produce palm oil, which is not exactly a small number. Except the total amount of land used globally for oil crop production is around 330 million hectares, meaning palm oil accounts for less than 10 per cent. Yet it produces 36 per cent of the oil that is consumed.[27] Compared to other vegetable oils, palm oil has a much higher yield. When compared to sunflower and rapeseed, the second and third highest-yield oil crops, palm oil is around four times higher in terms of its yield efficiency. Compared to coconut oil, it is more than ten times higher.[28] Ironically, although palm oil has such a bad rep, if the world's current demand for vegetable oil came only from palm oil, we would actually use four times less land than we currently do.[29]

Now is this all to say that we shouldn't worry about palm oil? Absolutely not. Palm-oil production is a leading driver of deforestation, and due to the areas of the world where it is grown, it poses a direct threat to ecosystems rich in biodiversity as well as important carbon stores.

While the situation is complicated and maybe not as simple

as we have been led to believe, there is still undeniably a serious problem with palm oil that needs to be addressed. So how do we do this?

In many parts of the world, the best action that could be taken to reduce the environmental impact of palm oil has nothing to do with food. The EU is the third biggest importer of palm oil, behind India and China. Two-thirds of the palm oil imported into the EU is converted into biofuels, which are used in transportation, electricity generation and heating.[30] However, research has shown that when land-use changes are factored in, using palm oil as a biofuel actually produces more greenhouse gas emissions than using a petrol car.[31] The good news is that the EU has voted to phase out palm oil use in biofuels by 2030.

The next solution would be to stop using palm oil in animal feed. A report carried out for the UK government showed that 23 per cent of palm oil imported into the UK is used for this purpose.[32]

And another solution would of course be to reduce the amount of palm oil that we consume in the foods we buy and the cosmetics and toiletries that we use. In regards to food, palm oil is mostly used in processed foods, especially things like biscuits. In fact, it's estimated that biscuits use more than 20 per cent of the imported palm oil in the UK.[33] If we eliminated its uses for things that we don't need, like biofuel and animal feed, and then dunked fewer biscuits in our tea, we would get significantly closer to making palm oil more sustainable.

If deforestation and species extinction are issues you are concerned about, then the bottom line is that the leading driver of deforestation[34], habitat loss[35] and species extinction[36] globally is animal farming. So palm oil being a plant-based product is not a reason to reject veganism.

'AVOCADOS ARE BAD FOR THE ENVIRONMENT'

Avocados are the punching bag of the fruit world. But do they deserve it? And even if they do, what does that mean for veganism? After all, the vast majority of avocados are not consumed by vegans. In other words, non-vegans drive up the demand, which increases the supply and then cite a problem they have contributed most towards to criticise a so-called vegan foodstuff.

But, from an environmental perspective, there are certainly some concerns around the production and consumption of avocados, with deforestation, water use and food miles often cited as being the main ones. From a deforestation perspective, there are serious issues about land clearance in Mexico. It is estimated that 30–40 per cent of recent deforestation in the region of Michoacán, which is where more than three-quarters of Mexico's avocado production occurs, is because of avocado farming. However, purchasing avocados from other countries like Spain and Israel avoids perpetuating the deforestation happening in Mexico, as well as helping to alleviate the reported issues of cartels vying for control of avocado production in certain regions of the country due to the significant value of the market. And when it comes to water use, 95 per cent of Spain's avocado production and 99 per cent of Israel's avocado crop is reliant on drip irrigation,[37] which can reduce a farm's water consumption by as much as 60 per cent.[38]

From a food miles perspective, people make the false assumption that avocados are flown to countries around the world, but this is not the case. Avocados are generally transported via boats, and when an avocado is shipped from South or Central America to Europe, only around 8 per cent of its footprint is from the transportation.[39]

In fact, per kilogram of edible food, avocados generate one third of the emissions of chicken, one-quarter of the emissions of pork, and one-twentieth of the emissions of beef.[40] Plus, because avocados grow on trees, they are also responsible for carbon sequestration, as the trees draw down carbon dioxide from the atmosphere.

However, the impact of avocados is often reported in ways that skew our perception of them in relation to other foods. An article in the *Guardian*, for example, had to be amended because the journalist hadn't specified that avocados have a large carbon footprint when compared to other fruits and had instead made the suggestion that they have a large carbon footprint when compared to other foods in general. A fact about water use, in which the journalist stated that one avocado needs 320 litres of water, also had to have an amendment attached to it, as this figure is an extreme example and is not typical.[41] This kind of reporting makes it really difficult for a consumer to make an informed judgement.

Similarly, without direct points of comparison, many of these figures are more or less worthless. While avocados might have higher carbon emissions and water usage than other fruits per kilogram, if those fruits were being compared based on their micronutrient content, protein or fatty acids content, avocados would rank far better, so fewer are needed to give the same nutritional value.

With all of that said, eating less avocados is not necessarily a bad idea, whether you are vegan or not, as a reduction in global demand would also help to ease the intensification of avocado farming. However, boycotting avocados altogether is not necessary and doing it for environmental reasons while continuing to buy animal products simply doesn't make sense.

'WE COULDN'T FEED EVERYONE ON A PLANT-BASED DIET'

This argument stems from the idea that if everyone were to stop eating animal products, we would need to grow more plants, meaning we would need more land. However, this isn't true.

The issue of land use for agriculture was analysed as part of a huge review carried out by researchers at the University of Oxford, which is considered the most comprehensive analysis ever conducted into the relationship between farming and the environment.

So what did they discover?

When looking at agricultural land globally, 83 per cent of all farmland is used to produce animal products, yet it only provides us with 37 per cent of the protein consumed globally and 18 per cent of the calories.[42] In other words, the production of plants for human consumption is vastly more efficient and means we can produce more food using less land. In the UK, 85 per cent of agricultural land is for farming animals,[43] which is just under half of the entire landmass of the UK.[44] Not to mention, the animal farming industries in the UK import more than 70 per cent of the main feedstuffs they use.[45] In the USA, 41 per cent of land in the 48 contiguous states is used for farming animals, compared to just 4 per cent that is used to produce plants directly for human consumption.[46]

So what would happen if we changed how we produce food?

If the land in the USA that is currently used to grow feed for animals was instead used to grow fruits, vegetables and pulses for human consumption, an additional 350 million people could be fed compared to what that area of land currently produces.[47] In the UK, 200 million people could be fed with a plant-based food system.[48] Food security and the self-sufficiency of the UK food sector is a concern close to the heart of the NFU. They are staunch advocates for UK agriculture, with the president, Minette Batters, stating that 'over the past few years, not enough importance has been placed on Britain's food production'.[49]

Martin Kennedy, the president of the Scottish NFU also expressed his concern about this issue to me during a debate we had together on Scottish television. He said, 'Do you realise that we're only 60 per cent sufficient for food here in the UK? Food security is right at the top of so many people's agenda right now.'

Well, Minette, Martin and the NFU in general will be delighted to know that if the UK repurposed the cropland it currently uses to grow feed for animals, it would easily produce enough food to be self-sufficient. In fact, the UK would be able to produce more protein and calories than are needed to feed the population.[50] Astoundingly, instead of embracing this fact, the NFU actually want to increase the amount of cattle the UK farms.[51]

If growing plants is more efficient than farming animals, what does that mean for the world? Well, according to the comprehensive review referenced earlier, a global shift to a plant-based diet would mean that we could feed every mouth on the planet and free up 76 per cent of current agricultural land.[52] This is an area of land equivalent in size to the whole of Australia, the whole of China, the whole of the EU and the whole of the USA combined. Let that sink in: an area of land that size no longer being needed for farming and we could feed every mouth on the planet. (To find out what we could do with all the freed-up land, see 'The Animal Farmer'.)

From a food production perspective, the real question should be: 'Can we feed every mouth on the planet if we don't stop eating animal products?' By the year 2050, if current trends continue, although our population will have only grown by about 20 per cent, our demand for animal-based foods will be 70 per cent higher compared to consumption levels in 2010, and our demand for ruminant meat in particular will be 88 per cent higher. The amount of land needed to meet this projected demand will be roughly equivalent in size to two Indias.[53]

Quite clearly something has to change.

'MONOCROPPING IS REALLY BAD FOR THE ENVIRONMENT'

Monocropping is an agricultural practice where an area of farmland is used for the production of only one crop for multiple seasons. Monocropping can be environmentally harmful because it can lead to the use of pesticides, the over-application of fertiliser and the implementation of poor soil management techniques, including tilling, which can impact soil health and fertility.

The logic goes that because vegans consume plants that come from monocrop farms, being vegan is bad. However, in the UK, the majority of cropland is used to produce animal feed.[54] In fact, only 34 per cent of the cereals grown in the UK are used for human food, compared to 50 per cent which are used to feed animals. In the USA, only 11 per cent of cereals grown are used as human food, while 42 per cent are used as animal feed.[55] So, a reduction in the production of animal feed in favour of crops for human consumption would actually help to alleviate the problems associated with monocropping.

The solution to monocropping is not then to eat animals – it's to improve how we produce plants. The impact of plant agriculture can be reduced with the introduction of techniques such as crop rotation, no tilling, and using cover crops and green manure.

And importantly for this conversation, not all monocrops are the same. Take silage, for example, which is preserved grass or maize that is most commonly used during the colder months in the UK when grazing animals are brought inside. Maize is reportedly the worst crop for soil erosion in the UK, with 75 per cent of the maize fields sampled in the south-west of England found to have degraded structure.[56] Of the maize grown in the UK, around 33 per cent is used for biogas, about 5 per cent is used for human food and more than 61 per cent is used as animal feed.[57]

So we might think of monocropping for animal feed being solely related to factory farming, but this is not true. Even outdoor-grazed animals can be given monocrops that have been sprayed with pesticides, herbicides, fertilisers and slurry, and harvested in ways that cause soil degradation.

And grasslands are not in most cases a natural alternative. In fact, according to Nature Scot, Scotland's nature agency, 'Most farmland grasslands are fertilised – or "improved" – to make them more productive.'[58]

Grass-fed advocates would jump on this and claim that this is why we should switch to purely organic, pasture-based, 'regenerative' farming. On what land though? Animal farming already uses half of the land in the UK[59] and 41 per cent of the land in the contiguous USA.[60] In fact, the poster child of so-called 'regenerative' beef farming in the USA has been shown to use two and a half times more land than a conventional beef farm.[61] Plus, nearly 2 million kg of feed is brought in annually from other farms, with no evidence that this feed wasn't grown using monocrop systems,[62] meaning it's land footprint is actually even higher.

So, their proposed solution is to switch to a more land-intensive and less efficient form of beef farming, even though red meat production is already more land-intensive and less efficient than any other system of farming, especially plant farming. Which national parks would that involve losing? Which forests and woodlands would we cut down?

By farming on more land, we would further reduce biodiversity, and considering that we know undisturbed natural ecosystems contain more soil carbon than their agricultural counterparts, we would reduce our ability to capture and store carbon on that land as well[63] (for more on the topic of the environmental impact of grazing animals check out 'The Pseudoscientist').

Furthermore, this argument is a red herring, as nobody is saying plant-based agriculture is perfect or can't be improved, because it can. But even the least sustainable plant foods have been shown to have less of an environmental impact than the most sustainable animal-based foods, with the lowest-impact meat (organic pork) being responsible for more than eight times more climate damage than the highest-impact plant (oilseed).[64] So not only is plant-based farming currently better but it also has the ability to improve and become more sustainable as well.

Ultimately, monocropping is not an argument against veganism. It is simply an argument for veganism combined with a shift to a better form of plant-based farming (see 'The Animal Farmer').

'WE ARE ALREADY DEVELOPING SOLUTIONS FOR FARMING EMISSIONS, SUCH AS SEAWEED FOR CATTLE'

Given that the scientific evidence so strongly points towards an urgent need for us to tackle the issue of animal farming in order to protect the environment, a slew of proposed solutions has been put forward by those who would like to have their steak and eat it too.

One such proposal is for the introduction of feed additives, particularly seaweed. This has become one of the most spoken about solutions due to some of the methane reduction levels that it has been shown to provide. For example, one of the most heavily reported studies revealed an 82 per cent reduction in methane levels in beef cattle that had been given a species of red algae.[65] However, the study only included 21 male cattle. Even though this was an extremely small sample size, it is still more than other widely cited studies, which included 12 dairy cows and 20 beef cattle respectively.[66,67] And such reduction levels have not been matched in all studies, with the longest trial of red algae feed additives showing a much more modest 28 per cent reduction in methane produced by the 40 cattle used in the study.[68]

One of the most significant issues with this proposed idea is that the seaweed is an additive to cattle feed. In other words, it is an impractical, and perhaps even impossible, solution for grazing cattle, as an additive can't be easily added to grass. This is also compounded by the fact that cattle have been shown to refuse feed with the algae, as they don't like the taste. The alternative would be to raise cattle entirely in intensive or zero-grazing systems, where feed intake is completely controlled. However, as well as the obvious ethical implications of this type of farming, the industrial system of feed production is also responsible for the production of greenhouse gas emissions, something that seaweed additives would obviously do nothing to reduce.

Another issue with this so-called solution is that a review of the research on algae additives showed that there was evidence of ulceration, haemorrhaging and inflammation in the cattle's stomachs.[69] Additionally, there is currently no supply chain in place to provide the world's cattle industry with the seaweed, meaning that a global, large-scale, commercially viable aquaculture operation would need to be created from scratch. It would be so much simpler if we just grew plants directly for ourselves. Not only would we eliminate 100 per cent of the methane produced by cattle, but we would also eliminate the absurd inefficiency of growing plants to feed to animals.

The other notable proposal that has been mooted is the use of methane masks, which would be placed around the noses of cattle. However, this idea is still at the research and development stage and therefore a huge way from being fully realised. So not only have these masks yet to be produced and tested to prove they actually work, but they would need to be made commercially viable, mass-produced and rolled out by the hundreds of millions worldwide.

Simply put, we should not be holding out on theoretical technology to deal with an urgent issue that needs addressing now. Plus, it is unnecessarily dystopian and ethically dubious to fit animals with respiratory devices in the first place.

Most importantly of all, even if feed additives, methane masks or any other proposed idea were found to reduce methane levels in ruminant animals by some of the numbers the industry has touted, this would only mean tackling one part of the issue around animal farming. The production of animal products, and in particular cattle farming, would still be the biggest driver of land use, deforestation, biodiversity loss and species extinction. Plus, these 'solutions' wouldn't lead to cleaner rivers and reduced nitrogen pollution, or reduce other greenhouse-gas-related issues, such as the emissions produced by feed production or the animals' waste.

There is, however, one solution to all of these problems: stop farming animals. Not only would this tackle all of these issues at the same time, but it would allow us to reclaim the no longer needed farmland and instead use it to store carbon and increase biodiversity through rewilding.

The problem is concepts such as these lure us towards complacency by offering us a supposed solution. This incentivises us to keep consuming animal products and offers legislators easy and politically friendly options that play into the industry's hands. In reality, these ideas are obfuscating the problem and disincentivising us from pursuing the solution that would actually create the largest and most well-rounded benefit of all: veganism.

If there's one positive to take from all this, at least the industries themselves are acknowledging that methane is a significant problem and that something needs to change. It is my hope that consumers realise what that change really needs to be and that these proposals are just more examples of the cattle industry producing hot air.

'YOU SHOULD SUPPORT LOCAL FARMERS AND NOT BUY PLANT-FOODS SHIPPED FROM ABROAD'

The local argument is a strange one. At face value it makes sense, and it's something that we hear all the time. But when you dive a little deeper, it just doesn't add up at all. In fact, Hannah Ritchie, a data scientist and head of research at Our World in Data, calls advocating for local food 'one of the most misguided pieces of advice' when it comes to protecting the environment.[70]

But why?

The basic idea behind this argument is that transporting food produces greenhouse gasses. So if we can reduce the amount of food miles that means that our food is more sustainable. However, there's an obvious problem here. For this logic to be true, every system of farming would have to be the same in terms of its environmental impact, with the only differentiating factor being the distance the foods have travelled. But we know that this is not the case. Not all farming is equal – far from it.

Because of this, we have to look at what is called the life cycle of food. In other words, the impact that an item of food has from beginning to end. Fortunately for us, this research already exists.

Let's take red meat. Only 0.5 per cent of emissions from the production of beef is from the transportation. For lamb, it's only 2 per cent. In fact, for most food products, less than 10 per cent of emissions associated with that food are from the transportation, with those that have higher percentages for transportation doing so because the farming is less impactful overall.[71]

Only 6 per cent of the emissions from the average EU diet are from the food's transportation. However, when split between animal foods and plant foods, 83 per cent of the emissions in the average EU diet come from the animal products.[72]

A landmark study that analysed the real diets of 55,000 people in the UK found that vegan diets resulted in 75 per cent less emissions, water pollution and land use than diets in which more than 100g of meat a day was eaten. Vegan diets were also found to reduce water use by 54 per cent and cut the destruction of wildlife by 66 per cent.[73] The researchers noted that what is eaten is far more important than where it was produced.

In the USA, the research shows that food transport accounts for only 5 per cent of the food emissions of the average US household, and that switching from red meat and dairy to plant-based alternatives just one day a week would reduce greenhouse gas emissions more than having a diet with zero food miles.[74]

So while it's clear that food miles shouldn't be the most

important consideration when deciding what we should eat, what about if we're talking about the same food? It might seem logical to presume that buying a food produced closer to home is better than buying the same food produced further away, but the evidence shows that not even this is necessarily true. Take tomatoes, for example. Research has shown that it is more sustainable for people living in Sweden to buy tomatoes grown in southern Europe than homegrown ones. The reason being that tomatoes are grown in greenhouses in Sweden, which means that ten times as much energy is used producing them there than it is importing them from southern Europe, where the climate is more suited to their growth.[75] Lettuce consumption in the UK is a similar story. During the winter months, importing lettuce from Spain results in three to eight times fewer emissions than buying locally produced lettuce.[76]

Part of the confusion about local products is that there is a misconception around the way that food is transported. Only 0.16 per cent of food miles are air-freighted, which means that the overwhelming majority of food is transported via boats, road vehicles and trains.[77] Boats emit 50 times less carbon dioxide compared to planes per tonne-kilometre (a unit of measurement for the transport of one tonne over a distance of one kilometre).

But food exports and imports are an issue that the animal farming industry takes very seriously. As my previous sparring buddy Martin Kennedy stated, 'A lot of the vegan diets rely on a lot of imports and imports of products that we cannot produce here, and that has implications worldwide.' And it is because of this concern that animal farmers don't import feed. Oh, wait . . .

We're told to support local producers, but more often than not those local producers are importing feed from abroad. That's right, buy that local British bacon, just don't think about the soya imported from South America to feed the pigs. And don't forget to get your local milk from a cow who was fed imported maize. But hey, at least you're not drinking soya milk that was imported from France.

And if importing food is so bad, why do animal farmers export food? British meat producers export to more than 50 countries and British dairy processors export to more than 135 countries worldwide. But they don't plan on stopping there – the animal-farming industry is currently working on growing pork, beef and lamb exports to the USA.[78] The NFU president Minette Batters believes that 'it's important that we grow our exports around the world and find new markets'.[79]

Animal farmers keep touting the same rhetoric over and over again. Joe Stanley, a beef farmer and a county chair and regional board member for NFU, stated:

> There is a danger of casting a very sustainable British industry to the wall in the pursuit of well-meaning campaigns such as Veganuary and then we may find that we're importing food from other parts of the world which have much worse environmental records and much higher carbon footprints.[80]

It's incredibly frustrating that climate scientists and farmers are being pitched as two equal interlocutors when they're not. Farmers are not climate scientists. If we wouldn't trust oil executives to be impartial and fair about the impact of fossil fuels, why do we still act as if animal farmers will be impartial and fair about animal farming? Climate and data scientists spend years producing evidence-based analysis, and then when they publish their findings animal farmers will just deny the science is true and the media report what they say as if it is a valid rebuke.

And the thing is, animal farmers don't disagree with the UN about fossil fuels. They don't disagree with any of the other climate science. The NFU isn't saying that climate change isn't a big problem. They agree with it all, except the bit that criticises what they do.

'QUINOA CONSUMPTION HURTS PERUVIAN FARMERS'

For some reason, people seem to equate quinoa with veganism, when in fact anyone can eat it, and then attempt to criticise vegans by labelling quinoa consumption as being unethical. On a live debate I had about veganism on ITV with the feminist campaigner Julie Bindel, quinoa farming was inevitably brought up. She said: 'If you look at quinoa, the new trendy food for the white middle classes in this country and elsewhere in Western Europe, Peruvian people, for whom that was their staple, now find that it's tripled in price.'

But is this actually true? Well, yes and no.

The price of quinoa did triple between 2006 and 2013. However, whereas Julie stated that increasing quinoa prices was bad for people in the South American countries where it is grown, the evidence shows the opposite. Quinoa farmers themselves have challenged the narrative being put forward by some in the Western media. 'To me, quinoa ... is absolutely changing the lives of our regional community of people,' stated one quinoa farmer.[81] The Bolivian president at that time even stepped in, saying, 'It's not true that due to an increase in the price of quinoa, less and less is being consumed.'[82] He even said that there had been a threefold increase in the domestic consumption of quinoa.

Confused by what was being stated in Western countries versus the information that was coming from Andean nations, researchers from the University of Minnesota and Towson

University in Maryland decided to investigate. They looked through ten years' worth of data, covering the few years before quinoa prices increased and the boom itself. What did they discover? That there was a rise in living standards and welfare in Peru.[83]

They also discovered that quinoa isn't even considered a staple food in Peru. For example, in the Puno region of the country, where more than 80 per cent of the crop is grown, quinoa represents about 4 per cent of the household food budget. Interestingly, the consumption of quinoa actually increased during the boom. Outside of the Puno area, quinoa represents less than 0.5 per cent of the household food budget.[84] Put in perspective, a staple food like rice or wheat in India is often responsible for making up somewhere between 25 to 57 per cent of a region's household food budget.[85]

The truth is, rural Andeans benefitted the most from the quinoa boom, with it bringing more income to some of the poorest regions. The people who were negatively affected by the price increase were those who were roughly twice as affluent as the farmers growing the quinoa and benefiting from the boom. But because of the wealth they already had, they could still afford the quinoa anyway, especially as it wasn't even a significant part of their food budget to begin with. As one researcher put it, 'It's a really happy story. The poorest people got the gains.'[86]

'FOOD WASTE IS A BIGGER ISSUE THAN ANIMAL PRODUCT CONSUMPTION'

Food waste, which includes consumer waste and losses in the supply chain, is undeniably a real problem, with some estimates calculating that it equals 6 per cent of total global emissions. This means food waste is responsible for around three times the amount of emissions of the aviation industry.[87]

However, research has shown that a plant-based diet would cut food emissions by half,[88] which would mean reducing global emissions by as much as 17 per cent.[89]

But even if food waste were a bigger issue, what exactly would that prove?

The logic behind this argument is not only unsubstantiated, it also creates a false dichotomy. We don't have to choose between a plant-based diet and reducing food waste – we can do both at the same time. So instead of trying to invalidate a plant-based diet by arguing for something that isn't true and doesn't need to be compared, we can alternatively work on achieving both things at the same time. That way it's just a simple win-win.

THE ANIMAL FARMER

'YOU'VE NEVER BEEN ON A FARM,
SO WHAT DO YOU KNOW?'

This argument is often tied in with the idea that vegans are city-dwellers who have lost touch with the natural world and don't get their hands dirty, instead choosing to cast judgement on the humble country folk who are stewards and caretakers of the land. I have had this argument levelled at me often, even though I've visited many farms and I'm a co-founder of an animal sanctuary.

And even if it is true that many vegans have never visited a farm, what exactly does this prove? You don't need to have physically been to a location to intellectually understand what that place is like, and, similarly, you don't have to have personally endured something to be able to empathise with someone who has. I've never been kidnapped, but I can surmise that being kidnapped is an awful experience. I've also never been to the Sahara Desert, but I know that I would find it very hot.

The same is true of farming. You don't need to have artificially impregnated a cow yourself to be able to work out that forcing your arm inside the rectum of an animal, holding their uterus in place through the lining of the anus and then syringing semen into it is weird, disturbing and immoral.

On the one hand, farmers use this argument as a way to

undermine veganism by accusing people who have chosen this lifestyle of being ignorant, while on the other using the fact that most people in society haven't visited a farm to their advantage by showing idealised versions of farming that focus on aspects they want the public to think of when they imagine farm life. For example, clips and imagery of fields being grazed by dairy cows are exceptionally common. Less common are depictions of a cow being forcibly penetrated and impregnated, or of a newborn baby being taken away from their mother and forced into a solitary confinement pen, and I certainly wouldn't hold my breath for a photograph of a dairy cow hung upside down inside a slaughterhouse being proudly displayed in the milk aisle of a supermarket.

I'm all for people seeing what farming looks like – it's the farmers who aren't.

'ANIMAL FARMERS LOVE THEIR ANIMALS AND DO EVERYTHING IN THEIR BEST INTERESTS'

A huge part of this argument is based on the assumption that because farmers are reliant on 'their' animals for their livelihoods, they would never do anything bad to them. However, is the same not true of puppy mill owners? Are they not also reliant on the animals in their facilities for their livelihoods, and so does that mean that puppy mill owners must love 'their' puppies?

And herein lies the problem: the reason farmers have animals is to make money from them, which means that the interests of the animals will always be weighed up against the financial costs. It is for this reason that dairy farmers send dairy cows to slaughter when their milk production wanes and they are no longer profitable. Nobody could argue that it is in a cow's best interest to be slaughtered – it is instead in the farmer's best interest, as that way they can make more money and free up space to bring in a new dairy cow whose milk production is profitable.

Now I don't doubt that some animal farmers do form a connection with some of the animals on their farm, and I know from talking to farmers that some do even feel guilty or sad the day their animals are taken to slaughter. However, any pretence that farmers only have the best interests of their animals in mind is obviously shattered when they load them onto a slaughter truck and then either breed or buy more animals to take their place knowing full well that the slaughterhouse will await those animals too.

I have even spoken to farmers who have proudly told me that they didn't send a particular animal to slaughter because they loved them so much. This surely reveals that these farmers recognise that slaughter is not an appropriate way to demonstrate love.

A case that didn't have such a happy ending involved an eight-year-old boy called Dalton, who was taking part in a farming scheme in the USA that involves children raising animals for slaughter. On the day that Dalton's lamb was to be taken away from him, he began to cry. His mother took a photo of him sobbing and posted it on social media, declaring how proud she was of him. She even said, 'My son ran the race and finished regardless of his feelings and emotions.'[1] There clearly was love for the animal involved in this situation, but the little boy had to ignore those feelings in order to send the lamb to their death. This example speaks to the perpetual need for farmers to not allow feelings or emotions to get in their way.

Unfortunately, Dalton was placed in a position in which the pride his mother felt came as a result of him carrying out an act that his emotions were struggling to contend with. Perhaps if he had thought that his mum would also be proud of him if he didn't send the lamb to slaughter, then he could have listened to his feelings and also felt pride in himself and from his mum at the same time.

'WE HAVE THE BEST ANIMAL WELFARE STANDARDS IN THE WORLD'

The ironic thing about this statement is that it is heard all over the world. Farmers from all over the world, including the US, UK, EU and Australia have told me that their countries have the highest animal welfare standards. But while it is obvious that not everywhere can have the best welfare standards in the world, it is true that there are also places that do. The UK is one of those places. The USA, Canada and Australia, incidentally, are not. But it is all relative, and the true test of how we treat animals should not be based on how we compare to other countries, but instead on what the animal welfare regulations in each country allow. It's not a commendation of the UK per se if on paper its animal welfare laws are better than those in the USA.

Improvements in animal welfare don't generally come about because of the farming industry wanting to make changes, but are instead created through public pressure, NGO campaigning and even via the influence of the supermarkets. The farming industry usually stands against such farming reforms, but when they are introduced, they often use the reforms to promote their welfare standards. Take gestation crates, for example. Pig farmers originally opposed their ban but now state that the importing of pork products from pigs kept in gestation crates is 'a betrayal of UK pig producers and the high standards of production they are proud to adhere to'.[2]

The truth is, stating that we have the best animal welfare standards is an easy and memorable soundbite that often means that debating what happens to animals in the country where we live can be avoided. Case in point, during the previously mentioned televised debate I had with Martin Kennedy, the president of the NFU Scotland, what was said certainly raised my eyebrows.

First Martin made the claim that 'we would be locked up if we treated animals in the way we're actually treating people in society right now'. I pointed out that we kill 90 per cent of the pigs in the UK using gas chambers, cut the throats of animals and selectively breed them even when doing so causes organ failure – treatment that humans are thankfully spared from.

When asked by the show's host to respond, as what I was describing was horrific, Martin said, 'Yeah, it is absolutely horrific. This is not the case at all. When you look at the welfare that's done through abattoirs, that's carried out now, we are not the same as what happens maybe in some other parts of the world. That is not happening here.'

Incredulously, I replied that I was 'absolutely shocked to hear that the head of the Scottish National Farmers' Union could lie so brazenly and so boldly on television'. Because everything I described happens in the UK.

It was bewildering to me that he could make such a statement. However, it was also pretty damning beyond it being inaccurate. After all, it's easy to have industry-proofed soundbites at the ready, but it's much harder to defend what happens to animals when the actual details of it are exposed.

Incidentally, in response to me calling him out, Martin backtracked, stating, 'What I was arguing was that you were saying that this was a brutal way of slaughter. That is not the case.' As well as clearly not being what he was arguing, this was also outrageous. If Martin doesn't think gas chambers and throat-slitting are brutal, I dread to think what he does classify as such.

The truth that we have better standards on paper than other countries is used to hide a falsehood, which is that those standards mean that what we do to animals is moral and caring. Even the best standards in the world allow, among many other things, for animals to be forcibly impregnated, to be selectively bred, to have parts of their bodies chopped off and cut out, to have their babies taken from them, to be trapped in cages and concrete pens, to be forced into gas chambers and to have their throats cut.

It's no surprise that Martin lied and then hoped the conversation would move on. The truth is damning and it shatters the facade of high animal welfare standards. If we really cared about the welfare of animals, we wouldn't exploit them in the first place.

'FARMED ANIMALS ARE BRED WITH THE PURPOSE OF BEING FARMED ANIMALS'

Applied consistently, this argument would make doing whatever we want to an animal morally justified if we bred that animal with that intended purpose in mind, but of course this is not a standard that we actually apply. In reality, we often view those who cause premeditated harm to others as committing an even more reprehensible act, as they acted with forethought and planning.

Moreover, this argument completely ignores the perspective of the individual who is being harmed, erasing their interests and desires to such an extent that they are not even viewed as a victim; instead, the only relevant aspect of this conversation becomes what the oppressor wants to do.

For example, this logic would make abusing an animal moral if you bred them to abuse them but abusing a wild animal or a stray animal immoral because they weren't initially bred to be abused. However, both examples of abuse are immoral because of what it means for the animals who are being abused.

Simply put, bringing someone into existence because we want to harm them does not make harming them more justifiable as a consequence. It just shows what little regard and concern we have for them to begin with.

'FARMED ANIMALS SUFFER LESS THAN WILD ANIMALS'

In order to justify our treatment of farmed animals, sometimes people will attempt to compare what happens to them to what happens to animals in the wild. After all, farmed animals don't have to worry about food or water, are less likely to be victims of predation and may even die in more preferable ways. While it is not necessarily true that farmed animals live better lives, especially as the vast majority are factory-farmed, it is also the case that some do.

However, why would this make eating animal products ethical? In the same way wild animal suffering doesn't make harming a dog or a cat moral, it also doesn't make harming a pig, cow or chicken moral. By consuming farmed animals we're not alleviating wild animal suffering. The suffering of an animal in the wild will still occur, it's just that because of our choices another animal has been bred into the world to suffer as well. This argument merely raises the point that the wild is certainly no peaceable kingdom nor a moral template for us to replicate (more on that in 'The Naturalist'). However, what is important is that first, regardless of what happens in the wild, the farming of animals doesn't decrease net suffering, it sadly increases it. And second, we can't use suffering and harm that we don't create and have no control over to excuse the suffering and harm that we cause and are entirely morally culpable for.

'IF WE ALL WENT VEGAN, WHAT WOULD WE DO WITH ALL THE ANIMALS WE HAVE?'

This is an argument I hear a lot, and, to be fair, it seems like a rational one at first. Vegans don't eat animal products, so if we were all vegan, the billions of existing farm animals wouldn't need to be slaughtered. Releasing all of these animals into the wild obviously wouldn't be good for the environment and wouldn't necessarily even be good for many of the animals, who could well suffer more due to the way we have domesticated and selectively bred them.

While there are certain influencing factors, such as the use of subsidies, that means animal farming isn't simply a case of supply and demand, it is still an industry that works on the principle of catering to the buying habits of the general public. The farm animals alive now have therefore been bred to fulfil the demands that the consumer has set through their previous purchases. However, as the number of people abstaining from eating animal products increases, the number of animals being bred would decrease alongside the falling consumer demand. Farmers wouldn't breed animals into existence if they couldn't then sell their secretions or flesh. A scenario such as this would lead to a gradual reduction in the number of farmed animals over time, meaning that if we were ever to get to a point where animal farming became obsolete, there would either be no farmed animals left or a tiny fraction of the number that are around today.

Even if we were to accept the most extreme scenario, in which veganism was introduced overnight and there were billions of animals that we couldn't release and had no choice but to slaughter, that would still be preferable to continuing animal consumption, because these existing animals were going to be slaughtered anyway, and they wouldn't be replaced with more to take their place.

In other words, this argument, even if it is ultimately redundant because of how the shift to veganism would unfold, basically boils down to a misguided notion that we have to slaughter, eat and keep breeding these animals because if we didn't we would have to slaughter these animals. This is clearly an absurd position to take when we are the reason why these animals are alive in the first place. If we simply stopped breeding these animals into existence, we wouldn't need to worry about the impact they would have.

'WE NEED ANIMAL MANURE FOR FERTILISER'

We don't need animal manure to fertilise our crops – what we need is the nutrients in manure that help with fertilisation, primarily nitrogen, phosphorus and potassium. We currently use animal manure because we have so much of it and because it contains the nutrients we are after, but that certainly doesn't provide us with a reason to keep farming and slaughtering animals.

So what could we use instead?

First, there are synthetic fertilisers. These have revolutionised the production of food, allowing us to improve crop yields while using less land and destroying fewer ecosystems. It's estimated that if crop yields had remained at 1961 levels, we would require nearly three times as much farmland today to produce the same amount of crops we are currently growing. This means we are using 1.75 billion fewer hectares of land as a result,[3] which is an area of land equivalent in size to the USA and Brazil combined. Synthetic fertilisers are also important for food security, especially in developing nations where the income of farmers and the subsequent security of the food supply chain are both positively impacted by their use.

But synthetic fertilisers are not without their problems. They can run off farms and into surrounding environments causing eutrophication in streams and rivers and affecting biodiversity. They can emit greenhouse gasses into the atmosphere as well. However, most of the issues related to synthetic fertilisers are down to two main reasons: over-application or poorly timed application. Research has shown that nearly two-thirds of applied nitrogen, for example, is not used by crops.[4] This is because it is common for farmers to add more nitrogen than is actually needed or to apply it at a time when the risk of run-off is high, such as during periods of rainfall.

In the mid-twentieth century, affluent nations began to offer subsidies to farmers to use synthetic fertilisers, which in turn led to those fertilisers becoming cheap and reduced the incentive for farmers to use them in the most effective manner. This has led to the significant over-application we have today and has contributed to the idea that using synthetic fertilisers is inherently bad, which is not the case.

One of the best ways to address this issue is to make the price of fertilisers in middle and high-income nations more proportional to the return on agricultural products, in turn incentivising farmers to apply fertilisers in a more sustainable, efficient and ultimately cost-effective manner. Those subsidies could then be retargeted to incentivise other forms of plant production and food production more generally (more on that in just a moment).

Technology can also be used to improve the efficiency of synthetic fertilisers, allowing farmers to pinpoint exactly which areas of their farms are lacking in nutrients. And reductions in fertiliser application have been shown to be possible while also increasing crop yields. A massive, decade-long study that involved researchers working with 21 million farmers across China showed that while fertiliser application decreased by around one-sixth between 2005 and 2015, the average yields of wheat, rice and maize increased by 11 per cent.[5] This was achieved simply by the farmers being trained on how to implement better fertiliser-management practices.

It's also important to note that the spreading of animal manure in the form of slurry can also lead to run-off and result in greenhouse gasses being emitted. In fact, animal farming is the biggest polluter of UK rivers and the main culprit is dairy farming precisely because of the manure. Dairy cows in the UK alone produce 50 billion litres of manure a year.[6] While we have control over the production of synthetic fertilisers, raising animals means producing manure that we then have to store or spread.

Switching to a plant-based food system would naturally reduce the quantity of fertilisers used, organic or otherwise, because it would require less land for agriculture and would require us to produce fewer plants overall as well. And it's not as if we would have to only use synthetic fertilisers either. Veganic farming is a method of producing crops using organic and animal-free methods. For example, by keeping the soil covered with cover crops. These are plants that are planted to cover the soil and are not intended to be harvested. Instead, they protect the soil from erosion and help to improve soil health and fertility, increase water retention and availability, and avoid nutrient run-off. They also help to control diseases and weeds and increase biodiversity, including attracting pollinators. And they are in effect a form of green manure. The cover crop can be turned over and its nutrients transferred into the soil, increasing soil fertility and providing the nutrients necessary to grow crops for human consumption.

Crops such as legumes are referred to as nitrogen fixers, meaning they take nitrogen from the air, add it to the soil and make it available for plants. Using nitrogen fixers for green manure is therefore a great way to get nitrogen into the soil. And because plants such as legumes, which include foods like soya, lentils and beans, are also eaten by humans, this means that growing nitrogen-fixing plants for human consumption can actually reduce the need for additional fertilisation in general. Veganic farmers can also use mulch and compost, turning crop waste, residues and waste from other industries into substances that can increase soil health and fertility.

Ironically, the nitrogen found in animal manure that makes it suitable as a fertiliser originated from plants to begin with, so veganic farming is about removing the part of the process in which animals consume the plants and simply harnessing the plants to begin with. This has the added benefit of cutting out all of the methane that's released because of the digestive systems of these animals and all of the other negative aspects that come with animal farming.

In many ways, veganic farming is about putting into place practices that have been employed for thousands of years and were used long before we had access to copious amounts of animal faeces or synthetic fertilisers. For example, Native Americans utilised what is called companion planting, whereby they planted corn, beans and pumpkins/squashes together, as the plants complemented one another and created a symbiotic form of farming.

There are already developed farms and communities of

veganic growers in the UK, the USA and elsewhere. If we are ultimately to shift away from animal farming, we will naturally need to employ other forms of fertilisation, and the good news is these methods are already being implemented.

'METHANE FROM CATTLE IS PART OF A NATURAL CARBON CYCLE'

After the publication of the first edition of this book, I received an email from someone who stated that it was a 'tour de force' – although they then immediately clarified, 'a tour de force in the wrong way for you'.

This person's source of frustration stemmed from an omission they believed I had made in order to avoid discussing something that would negate many of the arguments I had put forward about the environmental impact of animal farming. The omission in question was the biogenic carbon cycle that grazing animals are a part of.

Now, truth be told, the reason I didn't include this argument initially is because in many ways the animal farming industry itself has shifted away from it. However, considering it is an argument that still circulates, and considering I have been accused of omitting it on purpose, it seems prudent to address it. So, what is the biogenic carbon cycle argument?

Proponents of this argument claim that grazing animals are part of a natural carbon dioxide recycling system and therefore exist in some idealistic closed-loop cycle. This is because plants sequester carbon dioxide, the ruminants in turn eat these plants, and methane (CH_4) is created in their digestive systems. The methane is emitted and then breaks down into CO_2 after around 12 years in the atmosphere, where it is then sequestered by plants and the cycle repeats, which means (according to those who use this argument) that the impact of grazing animals is climate-neutral. Part of this belief stems from the fact that biogenic emissions are not the same as fossil emissions, because the latter reintroduce carbon that has been locked away for up to millions of years. On the other hand, biogenic emissions are part of a flowing cycle.

This theory paints a very holistic and symbiotic picture. However, what it crucially overlooks is that the rearing of these animals means that methane is being introduced into the carbon cycle unnecessarily, and this methane is having a warming effect while it is in our atmosphere. And, crucially, just because the methane contains carbon that was in the atmosphere initially does not negate its impact.

Even the assumption that methane from fossil fuels is far worse than biogenic methane doesn't hold up to scrutiny. The UN's Intergovernmental Panel on Climate Change (IPCC) shows that over a 20-year timeframe, while fossil-fuel sources of methane are 82.5 times more potent than carbon dioxide, non-fossil-fuel sources are 80.8 times more potent.[7] In other words, both sources are almost equally impactful, and both sources contribute significantly to global warming. As the IPCC states, 'increasing numbers [of livestock is] directly linked with increasing CH_4 emissions'.[8]

So there we have it. I didn't initially omit the biogenic carbon cycle argument because it represented an Achilles' heel for me. On the contrary, it is just another flawed argument that further reinforces that animal farming is destructive, environmentally harmful, and we'd be better off without it. Here's hoping the person who emailed me finds this response to be a tour de force in the right way for me.

'IF WE REMOVED GRAZING ANIMALS, WHAT WOULD HAPPEN TO ALL THE MANAGED AREAS OF PASTURE LAND?'

Farmers frequently make the point that animals are grazed on areas of land not suitable for growing crops and as such will ask things like, 'What are we going to do with all these areas of managed land?"

The simple answer is that removing grazing animals from these areas of land would provide us with one of the best opportunities to reverse biodiversity loss and sequester carbon dioxide, as we could rewild it and return it to nature. This would mean reforesting areas that had previously been deforested to make way for sheep and cattle farming. Many of us have a romanticised idea of pasturelands, especially in places like Scotland. But the reality is that these areas were deforested and destroyed and then turned into grasslands for animals to graze.

If the question posed was, 'Should we remove cattle from deforested areas in South America and reforest these areas?' what would we say? It wouldn't even be a debate. So why don't we apply the same logic to areas of deforested land that are grazed in the countries where we live? The problem is, we've been bombarded with misleading sentiments about the countryside, and the historical nature of deforestation in places like the UK makes it easier for us to fall for the myth that our landscapes are healthy and abundant when they are not.

Nature in the UK is actually in crisis, with almost half of Britain's biodiversity having disappeared. Shockingly, the UK ranks in the bottom 10 per cent of nations in terms of biodiversity intactness.[9] And the reason this has happened is because we have created pasturelands and managed agricultural lands that heads of farming organisations and many farmers are not only proud of but are actively trying to maintain.

A large review of more than 100 studies found a direct link between grazing animals and biodiversity. In fact, when grazing animals are removed from the land, the number and diversity of almost all wild animal groups increases. The only group in which numbers fall when grazing cattle or sheep are removed from areas of land are those that eat dung. The stark reality is that grazing farmed animals leads to there being fewer wild mammals, birds, reptiles and insects on the land, and fewer fish in the rivers.[10]

So what about greenhouse gas emissions?

Grazing animals produce methane through the process of enteric fermentation, so by removing those animals from our food system we also remove the emissions they produce. In addition, the rewilding of the land no longer needed for farming allows us to draw down carbon dioxide from our atmosphere. This is because of photosynthesis, the process that plants use to grow. Basically, they draw down carbon dioxide from the atmosphere and use it to build up their organic biomasses (i.e. their roots, trunks, branches, leaves and everything else that makes up the plant).

As previously mentioned (see 'The Misinformed Environmentalist') a plant-based diet would allow us to free up 76 per cent of agricultural land, which is an area equivalent in size to the whole of Australia, the whole of China, the whole of the European Union and the whole of the USA combined. If we were to re-wild that land, it would allow us to remove 8.1 billion metric tonnes of carbon dioxide each year[11], equivalent to around 15 per cent of our current global emissions from all greenhouse gasses.[12] This is on top of the emissions reduction that would occur as a result of no longer farming animals, which would be as high as 17 per cent.

So removing animals from our farming system would be

beneficial for a multitude of reasons and would allow us to better utilise the land we have. And there is actually precedent for grazing farmers reforesting their land. Which brings me to the next argument.

'VEGANISM MEANS THAT ANIMAL FARMERS WILL LOSE THEIR LIVELIHOODS'

Obviously when talking about the removal of animal farming, we're in essence talking about the removal of animal farmers' jobs. This understandably leads us to the question of how they would earn their livelihoods.

In the early twentieth century, Henry Ford changed the world with the introduction of the Ford Model T. Not only did it pave the way for modern automobile transportation, but it also led to a plummet in the demand for horses now that they were no longer the preferred means of getting from A to B. This had a huge impact on oat farmers, as the biggest consumer of oats in the USA were horses. So what did the oat farmers do? They diversified. The market changed, so they changed to meet the market. In another example, declining numbers of smokers in the USA led to tobacco farmers diversifying, many of whom started growing chickpeas to meet the growing demand for hummus.[13] In other words, the farming industry has long been accustomed to adapting to a changing society, meeting consumer demands rather than deciding them.

A transition to a plant-based diet would lead to a similar situation for the farming industry, whereby changes in consumer habits would lead to the opportunity for diversification and adaptation. It always strikes me as strange when people say, 'But what about the dairy farmers?' as if they, or any food producer for that matter, have a fundamental right to be supported. And by being vegan we are not removing our support for farmers, we are just supporting other farmers instead. After all, what makes a pig farmer more entitled to consumer support than a legume farmer? Or a dairy farmer more entitled than an oat farmer? Ironically, when someone talks about supporting grass-fed farmers, nobody says, 'But what about the factory farmers?' However, if someone says they're only supporting plant farmers, the question then becomes, 'But what about cattle farmers?'

There are different ways for farmers to diversify, from simply repurposing the land they use to produce animal feed or graze animals, to producing crops for human consumption. They can even utilise their agricultural space in more inventive ways. Chicken farmers, for example, raise tens of thousands of chickens inside long barns. However, these barns are also suitable for growing foods like mushrooms or plants like hemp.

It is important to note, however, that a lot of land isn't suitable for growing plants, and even if it were, we don't want to turn over all our current farmland to plant production for humans anyway. This is where the precedent I referenced in the previous argument comes in.

At the beginning of the twentieth century, Costa Rica focused on expanding its agricultural output, encouraging deforestation to create land for farming. Between 1940 and 1987, the country lost as much as half of its forest cover. Recognising the serious environmental damage that was being done, the Costa Rican government stepped in and made it illegal to cut down forests without explicit approval from the authorities. They then introduced the Payments for Environmental Services (PES) programme.

Today, while deforestation rates continue to increase in Central and South America, Costa Rica has not only halted their deforestation but has actually reversed it. The PES programme, which is financed by a tax on fossil fuels and money that has been freed up by redirecting public funds away from things like cattle subsidies, supports landowners by funding them to reforest, conserve biodiversity and protect the natural world.

There is no reason why other nations can't do something similar. Animal farmers are already supported with billions of pounds, euros and dollars. In fact, without subsidies, bail-outs and other forms of government support, entire systems of animal farming would simply not be profitable.

So rather than funding landowners to farm animals on their land, we can fund them to produce environmental benefits, such as carbon capture, cleaning waterways, improving pollinator biodiversity and the list goes on. Better still, such systems don't require landowners to halt the production of food completely. Rewilding could also be combined with agroforestry or the production of human edible plants, meaning landowners could make additional income at the same time as being supported through environmental subsidies. Other forms of income could come from eco-tourism, such as camping and glamping.

Alternatively, landowners could use their land to produce renewable energy, such as solar farms. Governments can also offer buy-out schemes, where landowners who are looking to retire or want a change of career can sell their land. Often these buy-out schemes offer above market value. Retraining schemes can also be offered to landowners who are looking to get out of farming. There is even the potential for fishermen and women to still live off the sea without fishing, instead growing seaweed like kelp.

And these ideas aren't at odds with what many farmers want. A recent survey of Scottish farmers showed that almost two-thirds with rough grazing and permanent pasture would consider transitioning out of animal farming entirely and into carbon capture through rewilding if they were financially supported. The same survey showed that 80 per cent of respondents would consider changing their farming practices in order to meet changing consumer preferences for plant-based diets, and 86 per cent said they would consider changing their farming practices to help mitigate climate change, with the majority of the respondents stating they would consider it even if it included a reduction in farmed animal numbers. As one respondent stated, 'We know these changes are coming so we have to be willing to change too.'[14]

The issue isn't necessarily that the desires of farmers and the desires of ethical vegans and environmentalists aren't compatible. When talking about farmers, I try to stress that while I oppose what they do, I understand how someone ends up farming and how the practices within it have become normalised. Farming is often a generational job, with sons and daughters inheriting farms that belonged to their parents and their grandparents. Farmers also tend to live in rural communities where their friends and community members are often involved in the agricultural world too. Living in such communities and within such environments doesn't remove accountability or responsibility for one's actions, especially considering there are animal farmers who do change, but understanding the dynamics and pressures that are a part of these communities is also fundamentally important.

Animal farmers are arguably the most important piece of the puzzle, as they can bring about huge change – and this change can benefit them and their communities as well. This is why the main source of my ire is the farming unions, levy groups and lobby groups. They are the ones who can bridge the divide between climate scientists, ethical vegans, government officials, the general population and the farming community. If they advocated for schemes to be introduced to help farmers transition, then that would not only normalise the idea for other farmers but would show the government and the general public that there is a desire for change.

Unfortunately, as it stands, it is these organisations who are perpetuating the anti-science argument, perpetuating lies about the climate impact of animal farming and perpetuating lies about how farmed animals are treated. They're misleading the farming community by not being open and transparent with them about what needs to be done, and they're halting progress that could be being made right now for the benefit of all of us.

In the same way that the heads of the fossil fuel companies and their special-interest groups are not naive, the heads of the farming world are not naive or scientifically illiterate either. If you are constantly having to reject or attempt to argue against what the most respected and comprehensive scientific literature is showing and, in turn, constantly disagreeing with the most respected scientific and climate organisations in the world, eventually you have to question why you are always at odds with what they're saying or accept that you have an ulterior motive.

At the end of the day, animal farmers don't have to be left behind in a transition to a plant-based food system – in fact, they are an instrumental part of the change that needs to happen.

THE PSEUDOSCIENTIST

'THE EMISSIONS FROM ANIMAL FARMING ARE NOT THE BIGGEST PROBLEM'

At the 2022 Wimbledon BookFest I had a debate with an anti-vegan author who began her piece by arguing that animal products have been unfairly maligned. To do this, she presented an infographic showing that animal farming is not the biggest source of emissions in the UK (although this of course applies globally as well). Sadly, this wasn't the first time I had heard this argument.

When it was my time to respond, I agreed with her before countering that something doesn't need to be the worst thing to be a concern or an issue that needs to be fixed. For example, the UK is reportedly responsible for around 1 per cent of the world's greenhouse gas emissions[1] – does that mean we don't need to worry about our impact or change anything because other countries are worse than us?

It should also be noted that in countries such as Ireland and New Zealand, animal agriculture is the leading emitter of greenhouse gasses, meaning that using this author's logic, eliminating animal farming should be their number-one priority. Funnily enough, I'm yet to hear the logic of this argument used for those countries.

However, there was another fundamental problem with her infographic – it didn't include the emissions produced by the parts of the UK's food supply that come from abroad. The animal farming industry is able to downplay its impact simply by counting emissions at home, which obviously completely overlooks the global nature of food production (this issue is covered in more detail in 'The Misinformed Environmentalist').

The figures for crop farming are often all grouped together as well, meaning that the production of animal feed is tallied with the production of plants for human consumption. This leads to the calculations for animal farming sometimes not including the emissions from feed production or the land needed to grow it. This obviously creates the impression that animal farming is not as bad as it really is.

During the debate, my opponent also referenced a 2017 study that just so happens to be one of the most cited studies by pro-meat advocates.[2] This piece of research calculated that if everyone in the USA went vegan, it would only lead to a 2.6 per cent reduction of greenhouse gas emissions and suggested the environmental impact of veganism would therefore be minimal. If we take this calculation at face value, 2.6 per cent of US emissions is the same amount of total emissions produced by Sweden, Switzerland and Portugal combined,[3] so even this supposedly small reduction would have a significantly positive impact. However, the authors of this paper made some very interesting decisions when they modelled what a vegan USA would look like in order to come up with the 2.6 per cent figure.

First, their calculations were based on people eating 4,700 calories a day, so around twice the amount we should be eating. Why would we need to eat twice as many calories in a vegan society? Vegans don't eat that many now. On top of that, the researchers also made the assumption that even though we would no longer be farming animals, we would keep producing all of the feed that we currently give to them and then consume it ourselves, leading to an unbalanced diet consisting mainly of corn. Why would we keep growing animal feed and then eat it ourselves if we went vegan? Wouldn't we repurpose some of that land to grow healthier foods, such as legumes, and rewild the rest?

They also included the burning of all the residues and waste from crop production in their calculations. Of course this would lead to increased emissions, but why wouldn't we simply compost the waste and use it to build up soil fertility, or use it as biofuel, or for one of the hundreds of other things it could be used for?

When you look at who the authors of the report were, it starts to make a bit more sense. Rather than respected climate scientists, one of them was a dairy industry research scientist and the other an assistant professor of animal and poultry science. This is the difference between good science and bad science. Authors of good science change their behaviours based on their findings; authors of bad science change their findings based on their behaviours. Unfortunately, this study, one of the most egregious and disgraceful pieces of food-related climate science published so far, is bad science.

Even if animal farming is not the number-one emitter of greenhouse gas emissions in certain countries or indeed even globally, it is the number-one driver of things like land use, deforestation, habitat destruction, species extinction and, in the UK at least, river pollution. So, if we are ranking issues on the leaderboard of total annihilation, animal farming certainly doesn't come out well.

The scientific consensus is that animal farming is a quantifiable concern. Pointing the finger elsewhere is not going to change that and, even though it might not be the biggest emitter, we certainly can't ignore it. A study produced by the University of Oxford showed that even if the emissions from every industry except for agriculture stopped immediately, the emissions from agriculture alone would take us past the 1.5°C threshold and make it difficult to stay below 2°C as well.[4]

This was backed up by a 2023 report that came to the same conclusion, stating that food alone could add nearly 1°C to global temperatures by the end of the century, with 75 per cent of this warming being due to methane-emitting foods – red meat, dairy and rice in particular.[5] Alarmingly, the study worked on the assumption that dietary patterns would remain constant through to the end of the century. But as the authors pointed out, 'the demand for ruminant meat is expected to increase by ~90% by 2050, while the consumption of all animal products is projected to grow by ~70%'.[6]

In other words, we don't actually have a choice – we absolutely must reform and change our food system before it's too late.

'ANIMALS ARE ACTUALLY EATING THE RESIDUES AND WASTE OF CROPS GROWN FOR HUMANS'

Let's apply some perspective to this. We slaughter around 85 billion land animals every year, not to mention hundreds of billions of farmed fish, who are sometimes fed crops as well. Plus the figure for land animals doesn't even include the billions of them that die on farms or on the way to slaughterhouses. On the flip side, there are around 8 billion humans.

The argument that these animals are only eating waste rests on the idea that not only are humans eating the food that is grown for them, but they are also eating and using all of the soya, maize, barley, rapeseed and other feed crops, with those tens of billions of animals subsisting on the waste from these crops. Obviously this is untrue. Yes, animal feed does include crop residues and waste from crop production, but it is not exclusively what makes up the feed we give to them, hence why as much as 75 per cent of cropland in the USA is for animal feed.[7]

So where does this idea come from? Well, one of the main crops quoted when it comes to animal feed being a waste product is soya. At the previously mentioned Wimbledon BookFest, the person I was debating levelled this argument against me, claiming that the soya used as animal feed is actually a by-product from the soya grown because of human demand.

The logic of this is derived from the market value of soybean oil and soybean meal. Soybean oil is actually worth twice as much per kilogram as soybean meal, so considering that it is the meal that is mainly used as animal feed that would suggest that the main economic driver of soya demand is actually the oil. But here's the important detail overlooked by those who make this argument. When you split a soybean into oil and meal, the meal makes up 80 per cent of the weight of the soybean. In other words, while it is worth more per kilogram, only about a third of the value of a crushed soybean is the oil. So this means that even though the oil is twice as valuable as the meal in terms of weight, the meal is worth twice as much as the oil when it comes to the soybean itself.

Beyond all of this, we just need to look at the rise of soya farming and compare it to animal farming to see the relationship between the two. Between 1990 and 2019, the global demand for soybeans grew from around 99 million tonnes to more than 329 million tonnes.[8] However, the global demand for human food products like tofu and soya milk grew by only around 3 million tonnes.

So what accounts for a tripling in soybean production between these two dates? Well, during the same time period, global poultry production grew from 41 million tonnes to more than 132 million tonnes. Who are the main consumers of soya globally? Birds raised for meat. Perhaps it's a mere coincidence that there was a tripling of the main consumers of soya, not to mention a near doubling in the amount of meat produced more generally, during the same time that soya production tripled.

So yes, animals do eat some crop residues and waste products. But no, this doesn't make up all of their diets, and what pro-meat advocates sometimes label as being waste is certainly curious to say the least.

'THE WAY THAT METHANE'S IMPACT IS CALCULATED IS OUTDATED AND UNFAIR'

Methane is one of the most important greenhouse gasses that we need to address, and this is for two main reasons. The first is due to its global warming potential (GWP), which basically means the heat absorbed by a greenhouse gas in the atmosphere as a multiple of the heat that would be trapped by the same mass of carbon dioxide. While carbon dioxide has a GWP of 1, methane has a GWP of 28 over 100 years and 84 over 20 years.[9] This is because methane is a more potent greenhouse gas and, compared to the same mass of carbon dioxide, absorbs more heat. But why are there two different GWP numbers for over 20 years and over 100 years?

The second reason comes down to the life cycle of different greenhouse gasses. Carbon dioxide is a stubborn greenhouse gas – it's like the person who doesn't get the hint the party's over and they should be going. Basically, once it's there, it's pretty hard to get rid of. In fact, it can stick around for hundreds of years. Methane, on the other hand, is the opposite. It actually breaks down in our atmosphere after around 12 years.[10]

The stubbornness of carbon dioxide leads to a compounding effect, whereby atmospheric levels keep rising and stacking up. In contrast, because of methane's short life cycle, it can actually reach something more akin to a plateau, whereby the amount of methane going into our atmosphere is levelled off by the methane breaking down. At this point, if methane emissions remain stable, the methane being produced does not contribute to any further temperature increases.

The discrepancies between the two gasses and the many different variables that need to be taken into consideration when comparing them make it hard to calculate the true relative impact of carbon dioxide versus methane. In particular, the conventional GWP metric fails to properly account for these variables, which has led to the creation of something called GWP*. This metric attempts to more accurately reflect how the life cycle of gasses changes their impacts over time by attempting to find the temperature responses of different gasses and using that as the point of comparison.

Animal farmers have jumped on GWP* and used it to claim that they are being unfairly demonised. Or, as the National Cattlemen's Beef Association put it, victims of an 'outrageous lie' that is being pushed by those seeking to sell plant-based alternatives.[11] In fact, they go one step further and claim that cattle ranching in the USA 'may not be contributing much at all to global warming'.[12] It sounds to me like they're emitting more hot gas than their cattle. But why do the meat industry like GWP*?

As we've seen, the metric looks at the relationship between emissions and temperature response, and because methane reaches a plateau point, this means that farmers can make the claim that they are not contributing to global warming as a result of the temperature no longer increasing because of their emissions. For example, let's say you're a cattle farmer in the UK. The farm has been in your family for generations and has always had a stable herd of around 100 cattle. When you come to take on that farm, the methane from those cattle might not be contributing to an increase in temperature because the increase has already happened and the amount of methane the animals produce is being offset by the natural breakdown of methane in the environment.

However, let's say that you're in a developing country and have just bought a farm and 50 cattle. Even though you are farming half as many cattle, you are seen as having a larger climate impact under GWP* because you have introduced more methane. This means farmers in agriculturally established countries can downplay their impact while placing the blame elsewhere. If this sounds unethical to you, that's because it is. As Joeri Rogelj, a climate researcher at Imperial College London, said, 'Because you polluted more in the past, you're allowed those emissions in the future. That is fundamentally unfair.'[13]

This is commonly referred to as grandfathering, a process in which pre-existing polluters get preferential treatment when it comes to environmental policy. Why should the farmer in the developing country be viewed more critically than the farmer in the UK, especially as atmospheric methane levels would decrease if either farmer stopped farming cattle? And this is the major flaw with the way that GWP* has been utilised by the meat and dairy industries. Even if stable methane emissions eventually lead to no further atmospheric methane increases, that doesn't change the fact that they have led to increases in both atmospheric methane and global warming in the first place. In fact, only carbon dioxide has contributed more than methane to the heating of our planet. US or UK cattle farmers can try and downplay their impact, but they have all contributed to the point we are at now, which is a near tripling of methane in the atmosphere over the past 250 years. It's nothing short of outrageous to claim that ruminant farming isn't a problem because of methane's short life cycle.

In addition to the issues with GWP*, drawing attention to the short life cycle of methane ultimately works against animal farmers and not in their favour. Back in 2021, Inger Andersen, who is the executive director of the United Nations Environment Programme, stated, 'Cutting methane is the strongest lever we have to slow climate change over the next 25 years and complements necessary efforts to reduce carbon dioxide.'[14]

It is the strongest lever because if we reduce methane emissions now, we can begin to slow global warming immediately. This makes tackling animal farming one of the best things we can do right now, because the benefit will be felt so quickly and because it buys us time to deal with the other issues we face, such as carbon dioxide.

And the best way to bring down atmospheric carbon dioxide levels right now? Rewilding and photosynthesis. So if we eliminate animal farming, we reduce methane, we slow down climate change and we give ourselves the best opportunity to utilise the natural world in our aim to reduce carbon dioxide levels as well.

Whichever way the global warming potential of methane is framed, the conclusion is the same: reducing methane levels leads to a limiting of global temperatures and helps to lessen the impact of climate change.

'MEAT COMPANIES ALL AROUND THE WORLD HAVE PLEDGED TO BECOME NET ZERO SO WE DON'T NEED TO WORRY'

The phrase 'net zero' can actually be quite misleading. For example, if we hear that animal farming is going to be net zero, many people believe this means the industry will produce no emissions. However, while net-zero strategies might partly involve reducing the total overall emissions being produced, net zero is essentially about offsetting emissions, meaning they are still being produced in the first place. In other words, if 30 per cent of emissions from the meat industry were being offset by reforesting, for example, the claim that the meat industry is producing 30 per cent fewer emissions would be misleading. Rather than producing fewer emissions, they would instead be offset by the carbon storage taking place through the reforesting. This is an important point, and one I'll come back to later.

Another issue with the concept of net zero is that it is an easy way to generate positive publicity with little oversight or responsibility. Take Meat & Livestock Australia (MLA), a lobbying organisation that is funded by levies on the animal farming industries in Australia. In 2017, MLA announced that the red meat industry would be net zero in Australia by 2030. Following this announcement, there was a lot of positive reporting and PR about it. However, something changed in 2024.

Not only have climate and agricultural scientists claimed that the target is unachievable, but MLA itself has now said that net zero is 'not necessarily something that needs to be met'.[15] In its place, MLA and the red meat industry are using the term 'climate neutral', which, incidentally, is the same term being used by the US cattle industry, which adopted a 'climate neutral' goal back in 2021.

'Climate neutral' and 'net zero' seem interchangeable. After all, the terminology is very similar, and it is not hard to imagine that the average consumer would perceive the two concepts in much the same way. However, 'climate neutral' relates to the emissions impact of animal farming using the GWP* metric, which I addressed and debunked in the previous argument.

This approach is also appealing to governments, as these calculations for 'climate neutrality' could give them a way of appearing to meet their climate commitments but without the need to limit animal farming or advocate for people to eat fewer animal products. For example, internal documents released under a Freedom of Information request revealed that the Irish government Department of Agriculture, Food and the Marine has considered a switch to GWP*, as otherwise it will need to reduce herd numbers to meet its climate commitments.[16] This is especially concerning, as agriculture is the number one source of greenhouse gas emissions in Ireland, responsible for twice as many as the transportation industry and more than twice as many as the energy industries.[17]

In New Zealand, where more than half of the country's emissions come from agriculture, the animal farming industries are lobbying for the government to adopt GWP*. Scarily, using this metric, Fonterra, the largest dairy exporter in the world, could claim to be 'climate neutral' by 2030 by reducing just 17 per cent of its current emissions.[18] And in the UK, although they have stated that they will be net zero by 2040, the NFU have also announced their support for the adoption of GWP* to meet their climate goals, with the National Farmers' Union president stating that 'the NFU will continue to engage with DEFRA on how methane reductions are recorded'.[19] The organisation says that they will achieve net zero partly through measures such as the use of feed additives (which I addressed in 'The Misinformed Environmentalist archetype') and farmland carbon storage (which I address later in this archetype). However, the main components of their strategy are bioenergy production and bioenergy with carbon capture and storage (BECCS).

Bioenergy is produced by growing a crop or tree and converting the biomass to energy – for example, by burning it. BECCS is essentially the same, although the process also involves capturing the emitted carbon and storing it underground. Both have come under intense scrutiny. Burning biomass has been shown to initially increase carbon dioxide compared to fossil fuels,[20] and BECCS is highly controversial, with serious question marks around its environmental benefits. Written evidence submitted to the UK government by Greenpeace warned that 'any net zero climate action plan that relies on BECCS is not a good plan'.[21]

In the UK, Drax power plant, formerly run on coal, now mainly produces bioenergy. Drax also started trialling BECCS technologies back in 2019. In 2022, BBC *Panorama* revealed that Drax had been acquiring wood pellets to burn for bioenergy from natural forests in the Canadian province of British Columbia. Drax denied this was the case but said it would not apply for further logging licences in the province. In 2024, *Panorama* then followed up on their investigation and revealed that Drax was still getting wood from British Columbia, including from forests that are 'rare, at risk and irreplaceable'.[22] In response, Drax admitted that it had in fact taken wood from old-growth forests.

And yet despite all of this, Drax still claims to be sustainable because international carbon accounting rules state that greenhouse gas emissions from burning wood are counted in the country where the trees are felled as opposed to where they are burned. This is why Drax can claim to be sustainable, even though accounting for these emissions makes them the single largest emitter of CO_2 in the UK.[23]

But how does this relate to the NFU? In 2021, the NFU announced that they were partnering with Drax because the company is a 'world leader in sustainable bioenergy'.[24] And, as part of their net-zero plans, UK farmers would contribute to the production of bioenergy. At the same time, the president of the NFU said in relation to their net-zero strategy that they had no 'plan to make any cuts' or to change their 'levels of production', and in response to data showing that meat consumption in the UK has gone down stated that 'there are other parts of the world that are hungry for high-quality meat'.[25]

But hang on a minute – if the NFU has no plans to reduce the production of animal products in the UK but also wants to start growing more crops to be used for bioenergy, where will the extra land needed come from? Well, not to worry, because the NFU has highlighted 'large areas of suitable land potentially available for diversification'.[26] However, what land are they referring to? If they have no plans to reduce production, it can only be arable land currently used for crops (meaning that the shortfall would have to be met by importing more crops from abroad), land that is currently used for animals (meaning further increasing the intensification of animal farming), or converting more land for farming. This means the NFU has no plans to convert any meaningful areas of farmland into forest either.

This strategy not only goes against the climate science but even contradicts a report commissioned by the UK government establishing that we need to eat fewer animal products and convert agricultural land to natural landscapes to meet our climate targets.[27] In the lead-up to the report, the president of the NFU at the time stated she had met then Prime Minister Boris Johnson to tell him that farmers wanted to see a change in focus from nature recovery back to food production.[28] It's actually impressive how completely out of line the NFU is with the evidence, and ultimately the whole thing is a complete mockery of any claims they make regarding net zero, climate neutrality or whatever phrase they want to hide behind.

Beyond all of the misinformation and propaganda around 'net zero' or 'climate neutrality' is the simple truth that these labels are based on the goal of balancing or offsetting the emissions produced by animal farming, and yet we don't need to balance or offset these emissions, as they don't need to exist in the first place. Even if BECCS turns out to be a completely safe and effective means of storing carbon, this wouldn't validate the existence of animal farming, because it makes no sense to offset emissions that come from an industry that is entirely dispensable and the removal of which would provide huge benefits for us, animals and the planet.

'THE EVIDENCE SUPPORTING A PLANT-BASED DIET FOR HEALTH COMES FROM WEAK SCIENCE'

In the 1940s, studies began to show a link between smoking and lung cancer, something tobacco companies vehemently denied. As the number of these studies grew, the tobacco industry continued to deny their findings. Even after the US Surgeon General issued a report highlighting the dangers of smoking in 1964, the tobacco industry stated that 'no scientific proof, then, has been found to convict smoking as a hazard to health'.[29] This continued throughout the twentieth century. How is it possible that such a huge body of evidence showing a link between smoking and cancer was being discredited and opposed for so long?

Epidemiology is the study of health and disease in a population. It takes a group of people, looks at their lifestyle habits and disease rates, and draws out associations. However, correlation does not always equal causation. So even if population studies of smokers and non-smokers showed a correlation between smoking and lung cancer, that didn't necessarily prove that smoking caused lung cancer.

Why do I mention this? Well, because the limitations of epidemiology are used to discredit much of the nutrition science around plant-based diets and the consumption of animal products.

Randomised controlled trials are the gold standard in health research. They involve splitting participants into a control group and an experimental group and managing the variables to see what the effect of what's being tested is. This obviously works great for things like medicines, where placebos can be given to the control group and the study can be carried out in a controlled setting. However, it's much harder to carry out a study that does something similar with animal products and lasts for a decade or more. The same is true of smoking. You can analyse a population of 100 smokers who have been smoking for 20 years and see how many develop lung cancer, but to create a randomised controlled trial is not only much harder but also has ethical implications. This is why nutrition science primarily uses observational studies to establish the impact of certain foods on health.

One of the ways to avoid the anomalies, biases and false correlations that might arise is to look at a body of evidence. While one observational study isn't necessarily strong enough in and of itself to prove a significant link between what's being researched, hundreds of studies all showing the same thing becomes a lot harder to ignore. This is why the WHO classified processed meat as a class 1 carcinogen and red meat as a class 2 carcinogen – not because of one study but because of what 800 different studies were showing.

Plus criticism of epidemiology is self-defeating. After all, what is the solution? To not carry out thousands of cohort studies looking at millions of people over a span of decades because it's not perfect evidence? To ignore the entire canon of research that consistently is pointing out patterns of correlation? If that is the logic we are using, then we also invalidate all the observational data looking at smoking. This is exactly what the tobacco industry argued for decades, and it's the same argument that the animal farming industries make now.

Ironically, anti-vegan advocates are quick to criticise the thousands of studies showing the correlation between animal products and negative health outcomes, or the consumption of more plants and positive health outcomes, but as soon as an observational study comes out that paints veganism in a bad light, they promote it as the smoking gun that proves veganism is bad.

One of the most glaringly hypocritical examples of this comes from the author Jordan Peterson, who only eats red meat and salt and has referred to the evidence supporting moves towards a plant-based diet as being 'dubious'.[30] However, in defence of his carnivore diet, he has cited one paper, a study in which the researchers went on social media and asked around 2,000 people who self-identified as carnivores to fill out a questionnaire. Unsurprisingly, the results from people engaged in pro-carnivore communities were positive.

When discussing the research on the Joe Rogan podcast, Peterson said, 'It's the only scientific paper I've ever read where the surprise of the authors was evident in the manner in which they wrote.'[31] That's certainly an interesting interpretation considering what the authors said was, 'These findings must be interpreted cautiously in view of several major design limitations. Our survey ... did not objectively assess diet, nutrient status, health-related outcomes, or confounding health-associated behaviours ... [I]ndividuals who experienced adverse effects or lack of health benefits are likely to have abandoned the diet and would therefore not have been captured in this survey.'[32]

But it's the evidence for plant-based diets that is dubious ...

With all this being said, a number of quality randomised controlled trials have been conducted to analyse plant-based diets. For example, one involving patients at a doctor's surgery in New Zealand found that those placed on a whole-foods plant-based diet 'showed several improvements with chronic disease risk factors and quality of life'.[33]

Another study found that a plant-based diet can reduce dietary acid load, which has been linked with cardiovascular disease and type 2 diabetes.[34] And a large review of studies that included randomised controlled trials concluded that plant-based diets should be integrated more into national guidelines and even considered as the first options for reducing cholesterol and for patients with type-2 diabetes, as they were shown to be more effective than medications.[35]

The reality is, there is a huge body of evidence supporting a plant-based diet. This evidence is not only plentiful but is so convincing that the world's leading authorities on diet and nutrition concur that a plant-based diet is healthy. These same authorities are also advising us to eat fewer animal products.

'THE EVIDENCE ABOUT RED MEAT CAUSING HEART DISEASE HAS ACTUALLY BEEN DEBUNKED'

Red and processed meat has long been associated with having the potential to cause health problems, including diabetes, dementia, strokes, cancer and, of course, heart disease. However, in 2019 all of this was debunked and recommendations to reduce our consumption of red and processed meat were scrapped ... or at least that's what some commentators led us to believe.

Phillip Schofield and Holly Willoughby, who were two of the UK's most popular TV presenters at that time, celebrated the good news by eating bacon sandwiches on live television and chatting about the study.[36]

Phillip: Bacon is back on the menu!

[*Cheering*]

Philip: Yes, yes, yes, yes! Previous studies have told us that we should be cutting down on our red meat consumption. But a new study out today, which I think might have been actually carried out by the same people who did the first one, who just read the information wrong – they've looked at the same evidence and said they could not be sure that the risks were real . . . I'm just going to say, when you read this stuff, just think, *Ah, I can't be bothered.* Don't take any notice of them, because they all change it.

So that's that then. Case closed.

Except no, Phillip was completely wrong. To begin with, there was no 'first one'. There have been countless studies looking at the impact of red and processed meat over the years. For example, as previously mentioned, when the WHO declared processed meat a class 1 carcinogen and red meat a class 2 carcinogen, they got to those classifications by reviewing 800 different studies. And the researchers behind the review he was talking about[37] did not conduct a previous study and 'read the information wrong'. That is completely untrue and minimises all of the work that has been carried out on this issue. But it's not as if communicating the correct information is important when talking to millions of people about issues regarding their health.

A good place to start when judging the validity of the report mentioned by Schofield is to look at the previous work done by the people behind it. Bradley Johnston, the lead author, published another paper in 2017 looking into the evidence behind recommendations to limit sugar intake. The paper concluded that 'guidelines on dietary sugar do not meet criteria for trustworthy recommendations and are based on low-quality evidence'.[38]

Interestingly, the paper was funded by the International Life Science Institute (ILSI), a group that is in turn funded by McDonald's, Coca-Cola, PepsiCo and Mars, among others. Now why would these companies fund a study that contradicted the mainstream scientific consensus that we need to limit our sugar intake?

On top of that, Johnston had been hired as an ILSI consultant. ILSI have even been criticised for financing studies that cast doubt on the dangers of smoking, and were involved in contradicting the WHO's cancer agency when it came to the cancer classification of the weedkiller Roundup.[39] Gosh, a scientist who has received funding from and consulted for an organisation that has tried to downplay the health risks of sugar and the cancer risks of chemicals and smoking. I wonder what other incredibly powerful and well-funded industry that has been criticised for causing health risks and cancer would want to work with such a researcher?

And this was only the tip of the iceberg when it came to Johnston and his co-authors' conflicts of interests. Johnston has also received funding from and works for AgriLife, an organisation funded by the beef industry that specialises in promoting red meat. When the study was published, it didn't disclose Johnston's previous ties to the meat and food industry.[40]

But even if we ignore all of this, the findings of the study were swiftly criticised by the American Heart Association, the American Cancer Society, the American College of Cardiology and Harvard's T.H. Chan School of Public Health, among others. Walter Willett, a Harvard professor of nutrition, called it 'the most egregious abuse of data I've ever seen'.[41]

I also think it's important to note that even if eating animal products didn't cause any chronic diseases or health problems, that wouldn't in and of itself justify continuing to eat them. While I think people should be given the knowledge needed to take control of their own health, the key reason I think people should stop eating animal products is not because they can be unhealthy.

After all, the reason it's wrong to kick a dog is not because kicking a dog might hurt our foot. The same is true in regard to eating animals and their secretions. It is wrong not because of what it might do to us but because of what it does do to them. So although I find it morally reprehensible and deplorable that people's health is being compromised due to industry-funded research, the reason why eating animal products is wrong is because it violates the moral consideration that should be granted to non-human animals.

'VEGANISM CAUSES BRAIN ROT'

In 2008, a study was published looking at the rate of brain volume loss in elderly people. The study followed 107 people aged 61–87 for 5 years and found that a decrease in brain volume was greater among those with lower vitamin B12.[42]

This piece of research has subsequently been reported on by sections of the media, including Fox News who ran an article headlined 'VEGETABLE-ONLY DIET UPS RISK FOR BRAIN SHRINKAGE'[43] that stated, 'The study said those on a meat-free diet are six times more likely to suffer brain shrinkage, as the most vitamin B12 is found in meats, liver, fish and milk.'

Except that's not true. In fact, the words 'meat', 'vegan' and 'plant-based' don't appear once in the actual paper, and it has nothing to do with analysing meat eaters and vegans. The paper did show that those with the lowest B12 levels had six times more brain volume loss; however, there was no mention of diet, and there was nothing to indicate that the participants in the study didn't eat meat. So to then jump to the conclusion that people on meat-free diets are six times more likely to suffer brain shrinkage is an absurdly obtuse and agenda-driven framing of a study that made no such claim.

The next major argument about veganism and brain health centres around choline, an essential nutrient in our diet. Just like the previous argument about red meat, the fears around choline and brain health in vegans stem from text-book PR by the meat industry.

In 2019 (a blockbuster year for meat-industry disinformation), there was a huge media storm around choline, with headlines like 'MOMENTUM OF VEGANISM THREATENING BRAIN HEALTH OF THE UK, EXPERT WARNS'[44] and 'PLANT-BASED DIETS RISK "DUMBING DOWN" THE NEXT GENERATION, NUTRITIONIST WARNS'.[45]

So what was going on?

An editorial was published in the *British Medical Journal* by someone called Emma Derbyshire that warned we could be 'overlooking a potential choline crisis in the United Kingdom'.[46] She insisted that moving away from the consumption of eggs, milk and meat could have unintended consequences by reducing the amount of choline in our diets. However, this was at odds with the British Dietetic Association, the leading authority on nutrition in the UK, which stated, 'You absolutely can meet the requirements with a vegan or plant-based diet.'[47]

Funnily, while Derybshire's editorial featured a graph mainly showing choline levels in animal products with just a few plant-foods included (albeit towards the lower end of the scale), she failed to mention that wholegrains and legumes have similar choline levels to meat and fish, and wholegrains, legumes, breakfast cereals, vegetables and fruits have higher levels than milk.[48] She did, however, call milk a 'major dietary provider' of choline[49]. Even more confusingly, her graph derived its data from a paper that showed that plant-foods are comparable sources of dietary choline. This was a clear case of cherry picking, with the top end of the graph stacked with animal products to create the impression that plants have much lower choline levels.

Choline deficiency can cause fatty liver disease, so if vegans were at a higher risk because of their choline intake, there should be a correlation between veganism and increased rates of fatty liver disease. Not only is this not the case, but studies have shown that veganism is correlated with a decreased risk of fatty liver disease.[50] Too much choline, on the other hand, can lead to an increase in TMAO levels, which is purported to increase the risk of developing heart disease.

Derbyshire didn't declare any competing interests when the article was initially published. Revisions were later added to the article, stating that she 'consulted for and advised: The Meat Advisory Panel, Marlow Foods (Quorn), the Health Supplement Information Service and the British Egg Information Service, amongst others', by which time the editorial had been published, a press release sent out and the global media frenzy was underway. Derbyshire is a consultant and PR rep, hired by industries and companies to consult and advise them. Where is the scientific integrity? Where is the journalistic integrity? Where is the accountability?

Veganism is now falsely linked with brain shrinkage and brain health problems because of some misrepresented statements about B12 and a PR stunt about choline from an industry-hired consultant whose editorial contradicts what nutrition authorities state, cherry picks data and advocates for dietary measures that could actually increase people's risk of an early death. So, no, veganism doesn't cause brain rot, and the only thing dumbing people down are these sensationalist and agenda-driven headlines.

'VEGANS SHOULDN'T USE WORDS LIKE "BURGERS", "SAUSAGES" AND "MILK"'

We've all been there, wandering around the supermarket looking for cows' milk when we see a carton of oat milk and pick it up because it has the word 'milk' written on it, which means it must be cows' milk, right? We then get home, open it up, and realise that the carton we have bought is actually oat milk. If only there had been some way we could have known.

In a state of outrage, we contact a consumer protection organisation demanding that something has to change – after all, it was only last week that we bought a packet of meat-free burgers only to get home and find out that they were actually meat-free. Too many consumers are buying products and then getting home to find out that what it says on the label is actually what they've bought.

Sarcasm aside, I do find it curious that nobody ever complained about coconut milk being called milk until veganism began to take off. Before then consumers didn't need to be protected from the word 'milk' being used for something that didn't come from an animal, but now it's an issue all of a sudden.

What about fish fingers? Fish don't even have fingers, yet I guess a consumer has no choice but to assume that they must because the label has the word 'fingers' on it. Unless the reason the word is used is just so that consumers get an understanding of the product they're buying because it creates a visual connotation?

Which is precisely what plant-based food companies do as well by using the descriptive words they use. Because let's be honest, a burger patty is just a food item that goes in a burger, and the word 'patty' refers to the shape of the food, and a 'burger' is what you end up with once it's inside a bun. We don't have to make burger patties with flesh. Same with sausages. The word 'sausage' describes the shape of a food, not its contents. Where's the political lobbying demanding that Scottish square sausages be forced to change their name because sausages are cylindrical?

In the same way that we say 'square sausages' so that a consumer knows that the sausage they're buying is a square, plant-based companies call their sausages meat-free so that consumers know the sausage they're getting is meat-free. Basically, if we can have square sausages and the space-time continuum hasn't folded in on itself, then we can have meat-free sausages without worrying too. And don't even get me started on hot dogs.

This whole argument used by the meat lobby is so cynical, disingenuous and patronising, and the double standards and hypocrisy are plain for us all to see. And people are not being tricked by plant-based companies. One study looking at this issue stated that consumers were not confused by words associated with animals being used to describe plant products, and, in fact, omitting such words actually led to them being less clear about how those plant products were to be used.[51] So, attempting to ban these words from being used for plant-based foods actually increases consumer confusion, the very reason the animal farming industries claim they want them banned in the first place. Plus, the FDA ruled in 2023 that plant milks can be called 'milk' in the USA. One of the major factors behind their decision was a consumer survey.

However, even if we take what animal farming industries say at face value, which is that they just care about consumers, their argument is exceptionally patronising and demeaning. They're fundamentally insinuating that consumers are too uneducated, ill-informed and naive to be able to understand labels. Don't worry, consumers of little intelligence, the caring meat, dairy and egg industries are here to protect us from unclear labelling because we're all just too unintelligent to understand that oat milk is made from oats and not an animal who was forcibly impregnated, had her baby taken from her and is going to be bled to death in a slaughterhouse.

And if we really want to talk about misleading labels, animal products are the best place to start. The animal industries are the masters of misleading consumers, using words and phrases like 'high-welfare', 'free-range', 'cage-free', 'outdoor-raised', 'outdoor-reared', 'farm to fridge', 'humane', and the list goes on and on. Hello, pot – meet kettle.

'OAT MILK IS WORSE FOR THE ENVIRONMENT THAN DAIRY AND CAUSES YOUR BLOOD SUGAR TO SPIKE'

Considering the popularity of oat milk, it is perhaps no surprise that it has recently begun to face a variety of criticisms. But do these criticisms hold up to scrutiny?

Let's start with the environment. Outside of being criticised because it's farmed in monoculture systems (see 'The Misinformed Environmentalist'), one of the most prominent criticisms of oat milk is that when comparing its protein content with cows' milk, it is responsible for producing slightly higher carbon dioxide emissions per gram of protein: about 1.1kg per 10 grams of protein compared to about 0.9kg per 10 grams of protein for whole cows' milk.[52]

It should be noted, though, that when comparing the two on protein count, cows' milk uses three times more land.[53] If this land weren't being used to produce cows' milk, it could be rewilded, leading to carbon sequestration and increased biodiversity. This is referred to as the carbon opportunity cost, which is the amount of carbon that could be stored on the land if we decided to return it to a more natural state.

Globally, the dairy industry is responsible for 1.9 billion tonnes of carbon dioxide equivalents each year. However, when you factor in the carbon opportunity cost of the land dairy farming uses, the industries' emissions jump to 5.1 billion tonnes of carbon dioxide equivalents.[54] In other words, if we were to take into account the carbon opportunity cost of cows' milk, even in the case of protein, cows' milk is more damaging for the environment. And this doesn't factor in biodiversity loss, eutrophication, freshwater use, air pollution, and all of the other damaging impacts of dairy farming.

However, this argument is also a red herring because it assumes that oat milk is being consumed for its protein content, when that's not necessarily the case. Oat milk isn't viewed or marketed as a high-protein drink and is instead commonly used for things like coffee. A fairer comparison would be calcium levels, considering this is the main nutrient dairy is known for. Fortified plant-based milks have the same amount or higher levels of calcium per glass as cows' milk. So, from a calcium perspective, cows' milk is far worse environmentally. Plus, even if protein was the most important consideration and the environmental harm was worse for oats than dairy, the answer wouldn't be to drink cows' milk – it would be to drink soya milk and pea milk.

So, what about health? Outside of being criticised for containing seed oil (see 'The Pseudoscientist'), the other main criticism of oat milk is related to its impact on blood sugar. The glycaemic index is a measure of how much a food impacts blood sugars. Oat milk has a higher glycaemic index than dairy and other dairy alternatives such as soya and pea milk. Saying this, oat milk still only has a 'moderate' glycaemic index. For reference, sweet potatoes also have a 'moderate' glycaemic index.[55]

But the glycaemic index doesn't tell the whole story. Instead, a separate measure called the glycaemic load gives a more accurate picture of a food's real impact on your blood sugar, as it encompasses its complete effect by looking at how quickly it makes glucose enter the bloodstream and how much glucose per serving it can deliver.[56] This is important, as watermelon, for example, has a high glycaemic index, but a serving has so little carbohydrate that its glycaemic load is actually viewed as 'low'.[57]

Oatly Barista, often the main focus of anti-oat-milk criticism due to its popularity, actually has a 'low' glycaemic load when you factor in the serving size. In other words, using oat milk in coffee is not something to be concerned about, and even drinking a whole serving of oat milk on its own is not a concern for most people.[58]

With regards to oat milk, the head nutritionist at Zoe (a programme that has itself been criticised for focusing too much on blood sugar[59]) stated that 'if you don't have diabetes don't worry too much about sugar spikes; your body will likely be able to handle it'.[60] There is even less of a concern if you're using oat milk in a meal that also contains a good serving of fibre, healthy fat and protein.

The problem is by hyper-fixating on blood sugar and viewing this metric as the most important when it comes to determining the healthfulness of a foodstuff, we are oversimplifying things, which can lead to us eating foods that are more harmful overall. For example, the popular influencer Glucose Goddess recommends eating beef jerky alongside a bowl of breakfast cereal to reduce blood spiking. But is adding class 1 carcinogens, which processed meats are,[61] really an advisable way to make breakfast healthier? According to her advice, a three-egg omelette with ham and mozzarella cheese could be viewed as a better meal than oats with fruit or an açai bowl with nuts, even though her suggestion would mean replacing healthy fats, antioxidants and fibre with saturated fat and a carcinogen.

Avoiding refined carbs, refined sugars and foods such as cakes, doughnuts and sweets is obviously advisable for your health. However, attempting to demonise wholegrains and healthy complex carbohydrates because of glucose is reductive and harmful. A huge review in the *British Medical Journal* concluded that 'whole grain intake is associated with a reduced risk of coronary heart disease, cardiovascular disease, and total cancer, and mortality from all causes'.[62]

Even with regards to glycaemic control, another wide-ranging review showed that wholegrain intake lessened the risk of type 2 diabetes and reduced fasting blood glucose levels. It concluded by recommending wholegrains as a means of diabetes prevention.[63] And another study showed that oat consumption has a beneficial effect on glucose control and lipid profiles in type 2 diabetics.[64]

It's perfectly normal for our blood sugars to increase when we eat; in fact, this is exactly what's meant to happen. And instead of trying to flatten our blood sugar curves by eating processed meat or other harmful foods, we should eat a plant-based diet that centres around legumes, wholegrains, nuts, seeds, vegetables and fruits – the foods that are the healthiest for us.[65] [66]

It's disappointing that oat milk has sadly and unfairly become the focus of criticism from those looking to beat veganism with a meritless stick. However, when looking at the balance of evidence, we can clearly see those criticisms are unsubstantiated.

'GRAZING ANIMALS ARE GOOD FOR THE ENVIRONMENT'

Faced with the overwhelming evidence that anthropogenic carbon dioxide emissions are changing the planet we live on, the fossil fuel industries found themselves in a precarious position. Like the tobacco industry, they had been funding campaigns to create doubt and confusion for years. However, when the evidence reached a point that their denials became untenable, a different approach was called for. While the fossil fuel industry and climate change sceptic lobby groups are still vociferous in their rejection of the problem, another argument began to emerge.

What if carbon dioxide emissions were changing the climate but not in a negative way? Rather than refusing to admit something was happening, what if they reframed the discussion instead?

And so carbon dioxide was described as being positive, as it meant more 'food' for plants. However, focusing on carbon dioxide emissions in such a reductive and simplistic manner overlooks the bigger picture. After all, we're emitting far more than is being sequestered by plants, and our emissions are contributing to global warming. Sadly, this argument is still regularly used by climate change deniers, including members of the United States Congress. But a powerful lesson had been learned: if you reshape the narrative, you can confuse people into thinking something that is bad is actually good after all.

Which brings us to grazing animals.

The science is unambiguous: the grazing of ruminant animals is the biggest source of greenhouse gas emissions related to diet, it is the biggest user of land globally, the biggest cause of deforestation and habitat loss, and the biggest barrier to us maximising the carbon capture potential of our terrestrial surfaces. In short, these animal products are the most environmentally damaging food items we can buy. And yet, in recent years, some people have attempted to reshape this narrative and make us believe that grazing ruminant animals are not bad for the environment – they're actually essential.

One of the most influential advocates of this message is Allan Savory, who championed this argument in his TED Talk in which he claimed that grazing animals on half of the world's grasslands would allow us to absorb enough carbon dioxide to return the world's atmosphere to pre-industrial levels.[67] The theory is that grazing animals increase soil carbon by eating grass. As they graze, the grass they eat is then replaced with more grass that sequesters carbon, and the animals then defecate, with their manure adding carbon back into the soil as well. He said, 'I can think of almost nothing that offers more hope for our planet, for your children, and their children, and all of humanity.'

You would think such a lofty statement would have at least some scientific backing, and yet no scientific research supports Savory's claim about the removal of carbon dioxide at the levels he stated. To defend his position, he references 'people who understand far more about carbon than I do'. Although strangely those people don't seem to have any published data. Not exactly the overwhelming scientific evidence you might hope for or expect.

In fact, the science has comprehensively disproved his claims. A study in China found that livestock-free grasslands sequestered more carbon, with soil carbon levels being 157 per cent higher than grazed grasslands, and removing animals from scrublands in Australia resulted in net carbon sequestration.[68] One of the many rebuttals of Savory's position stated that his claims are 'not only unsupported by scientific information, but they are often in direct conflict with it'.[69]

The argument for grazing animals persists nonetheless, with some people suggesting that pasture-raised animals, while not necessarily able to sequester the levels Savory claims, are able to sequester enough carbon to make them a part of the solution to the climate crisis. Advocates of such a system argue that grazing animals can lead to the emissions from animal farming being completely offset by this seques-tration. But what does the science say?

An extensive and detailed two-year review called *Grazed and Confused* looked specifically at these arguments. It ana-lysed more than 300 sources and was conducted by an international team of researchers, including those from two of the top three agricultural institutions in the world. Their research showed that while certain grazing management sys-tems resulted in carbon being sequestered, even with the most generous estimations, grazing animals could only offset 20–60 per cent of the emissions that they produce in the first place and are therefore net contributors to the climate problem.[70]

On top of that, a report by the UK government stated: 'The extent to which soil carbon sequestration can offset agricultural emissions in the UK is uncertain at present but the best available science indicates that potential in the UK is limited.'[71] And this from a government department that regu-larly sides with animal farmers.

Plus, even if it were possible for animal farmers to seques-ter more emissions than they produce, after a few decades soils reach something called soil carbon equilibrium. At this point, any carbon being stored is matched by carbon that is being released and no emissions are being offset. This means that even if the claims made by grazing advocates were true, it wouldn't be a long-term solution to the problem, because at a certain point we would either have to contend with the emissions the animals were producing, create more agricultural land that the animals could graze on or stop the animal farming.

Unfortunately, the pro-grazing argument has caused a huge buzz within the agricultural world: General Mills, Unilever, Danone, PepsiCo, Cargill and Walmart, among others, are all getting behind regenerative agriculture. Worst of all, even JBS, the world's largest meat producer, responsible for things like Amazon deforestation and bribing politicians, are working on a programme focused on carbon sequestration. That's right, regenerative beef from one of the main companies behind Amazon deforestation and forest fires. I guess we have nothing to worry about after all.

Although words like 'holistic', 'regenerative' and 'restorative' sound lovely, there's a reason why the data doesn't support the farming of animals, and there's a reason why it is the food conglomerates and animal farming community who are the ones championing it and not the scientific community. Ultimately, the BS from these industries isn't just coming from the animals being farmed.

THE NATURALIST

'SOME ANIMALS EAT OTHER ANIMALS, SO WHY CAN'T I?'

While it is certainly true that animals in the wild eat other animals, is it really wise for us to use other animals as a model for our own actions and ethical stances? After all, other animals also rape and kill one another – are we morally justified to commit those acts just because other animals do?

What I find so interesting about this argument is that the only behavioural trait we actually seem to want to take from animals is their killing of one another for food. In other words, we already know that non-human animals shouldn't be viewed as moral icons.

However, I don't necessarily think that people who make this argument really want to be like other animals; instead, they seem to be insinuating that eating animals is simply the way of the world. Certain animals being eaten is a fact of life, so humans eating animals is an extension of this.

The question then becomes: why do other animals eat animals? Simply put, because they have to. A lion kills a wildebeest in order to survive. In other words, they do so because they have no choice – we, of course, do.

The argument is sometimes then brought round to that of animal omnivores, who could theoretically survive on a plant-based diet like we can. However, being vegan stems not just from the theoretical ability to survive without animal products but also the intellectual know-how to eat a balanced diet and the ability to make a principled decision about why it is beneficial to do so. Even if a bear could be vegan, they are not aware of this fact, and they also lack the moral agency to be able to rationalise the ethics of such a decision. Humans are undeniably different in many ways to other animals, most obviously in our intellectual capacity and ability to conceptualise ideas such as animal rights.

It is these differences that make the topic of veganism relevant and morally important to our species, and these differences that create a clear distinction between the actions of humans compared to those of other animals. Unlike them, not only are we able to be plant-based, but we are also able to rationalise the ethical importance of that choice.

'IF VEGANS WERE STRANDED ON A DESERT ISLAND, THEY WOULD EAT MEAT'

If someone were stranded on a desert island and their only option was to eat a pet, they would probably do it. But what does that prove? Simply that when placed in an extreme situation we are more likely to do something morally objectionable in order to survive. Even if a vegan would eat meat if they had no other choice, that doesn't morally justify eating animal products when we do have one.

Robinson Crusoe was no moral philosopher and nor should our morality be defined by what someone would do in such a situation. Plus, I'm not sure at what point I should be concerned. The number of times I've been asked what I'd do if I were shipwrecked on an island makes me think people know something that I don't. That said, with the number of vegans who have so far been stranded on this island, at least I can be confident that there's a really successful falafel and hummus pop-up there by now.

A hippy vegan commune filled with hummus, poorly made clothing and a distinct lack of deodorant sounds like what my family probably envisioned my life would be like when I told them I had gone vegan. Although, in their defence, I did make the change in 2015 when attitudes towards veganism were more derisive and when, in all fairness, there was probably a higher chance of getting stranded on a desert island than there was of finding good vegan cheese.

Today there is an abundance of good plant-based alternatives for all sorts of foods, including cheese, not to mention cruelty-free deodorants, and the likelihood of a vegan becoming stranded on an island is about as likely as a vegan never hearing this supposed rebuttal again. In short, it doesn't matter what a vegan would or wouldn't do in this hypothetical situation. All that matters is that when we have a choice to reduce the harm we cause to others, we should.

'EATING ANIMALS KEEPS US ATTACHED TO NATURE'

'I am part of nature, not above it. I choose to be within the food chain, not to stand outside of it. I think nature has given me a pretty good path to follow, just like it gave all the other creatures a path to follow to survive.'[1]

Many people fall into the fallacious trap of thinking that nature provides us with some form of moral justification for what we do to animals. The quote above by a farmer appeals to that very idea. I think it is completely beyond any sense of irony that people can make these kinds of claims while reaping the benefits of a modern society. How many people who claim they are a part of nature turn down life-saving medicines? How many people who claim they are a part of nature don't live in a house? How many people who claim they are a part of nature refuse to eat any food that hasn't been hunted or foraged? After all, agriculture isn't natural. Our food system is human-made. So when we raise and kill animals, that's not natural.

Plus, it's not surprising that this farmer is happy with the path they are on – they're not being farmed, mutilated and slaughtered. I'm sure if the roles were reversed they would feel differently; for example, if an alien race came to Earth who wanted to subjugate and eat us. I wonder how many naturalists would be willing to submit to such a scenario if it did occur.

Or what about cannibalism? It happens regularly in nature, so should it be considered moral for a human to kill and eat another human? If not, then we already recognise that what is natural is not necessarily what is moral or preferable for humans.

This touches on one of the more disturbing implications of this argument. It again subscribes to the philosophy that might is right – that what is moral is determined by strength and physical prowess. We are allowed to subjugate and dominate other species simply because we have the power to do so.

This idea can be applied to any situation of oppression. It is a common trope in misogynistic and white-supremacist groups, with the proponents of those ideas believing they have been granted superior traits by nature and that is why their bigotry is justified. It is simply a convenient way of trying to excuse the inexcusable and remove moral accountability from one's actions.

The idealisation of nature speaks to just how separate from nature we actually are. It is the fact that we are removed from it that affords us the privilege to romanticise it. I raised this point during a debate I had with a hunter for the media outlet LADbible, who stated that she had daydreams about living in caves and that it would be like a Disney movie. As well as pointing out that the reason she daydreamed about such an existence was because she didn't have to deal with the actual hardships of it, I also said that being a hunter would actually make her the villain of a Disney movie.

Nature is not a moral authority, and the natural world is not aspirational. It is a violent, cruel, merciless place where there is a good chance you'll die in a horrible and brutal way. The natural world is a product of evolution, through which decisions are not made according to moral virtue but simply as a result of natural selection increasing a species' ability to reproduce and therefore survive.

In the previously mentioned article, the farmer also said, 'When I was a vegetarian for about seven years, I began to feel that I had actually judged nature.' Acknowledging that nature is imperfect and there are things that happen within it that we shouldn't replicate is not a bad thing. We shouldn't treat nature as if it is a deity that shouldn't be questioned. It's simply a term used to describe a complex web of systems that creates and maintains life.

Even with regard to our treatment of animals, most of us agree that reducing suffering is preferable, a point the farmer also makes when she discusses trying to reduce stress in the animals she kills. But if we agree with that, then surely the pursuit of reducing suffering is more important than trying to replicate practices that occur in the natural world. And if reducing suffering is preferable, that leads us towards veganism and away from the natural world, which is filled with suffering.

Fundamentally, we've already taken the step to do something unnatural by farming in the first place, so we should be working to make that system of farming as ethical and sustainable as we can. Plus, wouldn't we much rather live in a world in which we empathise with the situations of others and actively attempt to minimise the harm that is caused? Especially considering that we would want others to empathise with what we were being forced to endure if we were the victims. If you want to feel close to nature, just go for a walk in a forest and leave the animals alone.

'EATING ANIMALS IS PART OF THE FOOD CHAIN'

Appeals to the food chain overlook what a food chain is about. Food chains don't exist in nature so that humans who want to eat animal flesh can do so and then claim that their needless slaughtering of someone else helps them stay in touch with the natural world that they don't actually want to be a part of. A food chain in nature exists to stabilise population sizes, whereas the food chain that we most often talk about is a human-made construct. And not only is agriculture a human-made system, but the food chain we are saying we are a part of is also the leading cause of things like species extinction, habitat loss and environmental collapse, the very things that natural food chains help to avoid.

Our consumption of animal products is literally achieving the opposite of what a food chain is supposed to do, and irrespective of that, we are not bound by the power structures of nature. We are not tied into some deterministic situation where we have no choice but to slaughter animals.

'EATING ANIMALS IS PART OF THE CIRCLE OF LIFE'

This is similar to the food-chain argument in the sense that it attempts to relinquish moral accountability for our actions by claiming that we are all entrapped by a form of pre-ordained fate. But the circle of life simply means that all who are born will one day die. And just because all living things die doesn't mean that we are justified to kill someone else, irrespective of whether or not death is ultimately unavoidable. We could use this logic to justify murder and the behaviour of serial killers. Sure, Jeffrey Dahmer killed all those people, but it's just the circle of life.

Similar to the food-chain argument, it is easy for us as humans to cite the circle of life as a justification for our consumption of animals – after all, we're not the ones being oppressed in this scenario. The irony is, we exempt ourselves from situations that would arise if our logic were used consistently. On the one hand, we use this argument to justify causing harm, suffering and death to others as being part of life, but on the other hand we create laws and social norms to try and avoid harm, suffering and premature death being inflicted upon us.

'VEGANS WANT US TO EAT LAB-GROWN MEAT AND MOVE AWAY FROM "NATURAL" FOODS'

No, we don't.

We want people to stop treating animals as disposable commodities and give them the moral consideration they deserve. If people want to eat lab-grown meat when it's available or eat processed meat alternatives, then that is entirely up to them. However, the narrative that vegans have an agenda to move people away from 'natural' or 'real' foods, as I've heard them being described, is not true at all. Most vegan advocates actually emphasise the importance of making sure a plant-based diet is a whole-foods one for health reasons.

And even though I am sure there will continue to be a rise in meat alternatives and foods that utilise novel and inventive production technologies, these are by no means required if you want to be a vegan.

Ironically, the reason foods such as meat alternatives are being produced is because of non-vegans. Cell-cultured meat is being developed so that people who eat animal products have the option to continue doing so but without harming animals. People often tell me, 'When lab-grown meat is available, I'll stop supporting animal farming,' to which my response is always that we shouldn't wait for food technology – we should just eat plants now.

And as discussed previously in this chapter, it's also ironic that pro-meat advocates promote natural foods when you consider that agriculture is fundamentally unnatural and involves practices such as selective breeding.

'HUNTING IS MORE SUSTAINABLE THAN VEGANISM'

Hunting is only viewed as sustainable because most people don't get their food that way. If they did, it would be unbelievably unsustainable. We would wipe out entire species of animals in a matter of days and would quickly run out of food sources.

Even when we were hunter-gatherers, we hunted to extinction entire genera, a term used to describe a group of animals or plants. During this time of species extinction, it is estimated that there were fewer than 10 million hunter-gatherer humans on the planet. There are now 8 billion of us. How could hunting ever be considered a sustainable alternative to animal farming?

In fact, the only reason our population ultimately grew is because we moved away from being hunter-gatherers and became farmers. The world is simply not able to support a situation where meat is derived from hunting.

When we talk about sustainable food systems, we're talking about precisely that – food systems. In other words, we need to find solutions that will allow us to feed people sustainably at the population level. Hunting clearly isn't able to do that, so it just doesn't work as a sustainable alternative.

Hunters also argue that they contribute to the funding of wildlife agencies and conservation. However, the most recent survey from the US government's Fish and Wildlife Service reveals that while hunters generated $26.2 billion in 2016, of which $800 million came from licences, stamps, tags and permits, wildlife watching activities resulted in an income of $75.9 billion, with $3.8 billion of that coming from memberships, dues and contributions.[2] So incentivising non-violent engagement with the natural world generates more money.

Arguably the most common argument that hunters use to defend their actions is to claim that they hunt because certain species like deer run the risk of overpopulating. While this would obviously discount the hunting of predators such as wolves and bears, who some hunters also kill, is hunting the best solution to this issue?

Part of the problem is how we perceive different species of animals. Stray dogs and cats are damaging to wildlife. They can run into traffic, they suffer from diseases and starvation, freeze to death, and so on. Should we hand out hunting licences so hunters can kill stray cats and dogs and stop their populations from growing at an exponential rate?

It would actually be illegal to hunt a stray cat or dog, even if you said you were doing so for conservation reasons. But what's the difference? If we can hunt some animals because of overpopulation risks, why can't we hunt other animals for the same reasons? Plus, stray dogs and cats are arguably the most sustainable form of meat. They're invasive species who don't positively benefit the urban environments they're found within, and they kill wildlife like birds and rodents. (I'm being facetious, of course.)

Instead of hunting these animals, spay and neuter programmes sterilise stray animals to stop them from procreating. We could carry out similar programmes with other animals, like deer, opting to sterilise them rather than kill them.

This practice of sedating wild animals is known as green hunting and has been used across parts of Africa, whereby animals such as elephants and rhinos are tranquillised so that veterinarian procedures can be carried out and they can be tagged and tracked. Green hunting allows for population control and monitoring but doesn't result in animals being killed. The use of immunocontraceptives has also been trialled in some places, with them being responsible for a 60 per cent decrease in the deer population over a five-year period in one area.[3]

The main focus for hunters isn't really overpopulation, though, which is why it's still a problem in many places. Hunting licences generate money, and people hunt because they enjoy doing so. If the main focus of hunters was to act in a way that best managed the overpopulation of deer and other species, they would specifically hunt female animals. The only issue being that they don't have big horns and aren't as impressive on someone's mantlepiece or in a photo on their social media.

It is also ironic to claim that the overpopulation of deer affects biodiversity when the biggest driver of biodiversity loss is animal farming. If a hunter hunts because they're worried about biodiversity but doesn't advocate for people to not eat farmed animals and to be plant-based then protecting biodiversity is not really their concern. And it's important to consider that animal exploitation is the primary reason why population imbalances occur in the first place. Predators are considered by some hunters to be big prizes, and have therefore been hunted for food and for fun. Animal farming has also contributed to their demise. This is because a wild predator doesn't differentiate between a fellow wild animal or a domesticated animal in a farm. In fact, a domesticated animal in a farm is a much easier target, so wild predators pose a direct threat to the profits of farmers. This has led to the decimation of wild predators.

The threat of wild predators to farm animals has also led to the creation of a USDA agency called Wildlife Services – this name makes it sound as if the agency exists to be in the service of wildlife, but sadly the opposite is true. Wildlife Services are responsible for killing more than 2 million wild animals every year, with some years the number being as high as 5 million, and it costs the taxpayer more than $100 million a year for this 'service'.

And their methods are indiscriminate, which has led to people's dogs being caught in traps, and a 14-year-old boy was also covered in cyanide when he and his dog triggered an M-44 (also known as a cyanide bomb) that had been placed 350 feet away from his house because of a request from a sheep rancher. While the boy survived after being rushed to the emergency room, he watched his dog convulse and die. The American Sheep Industry Association has referred to M-44 cyanide bombs as being a 'critical tool' that has a 'proven track record of protecting livestock and the environment'.[4]

It seems peculiar that on the one hand we are told that the deer population is out of control and on the other hand every year more and more predators who would naturally keep them in check are being killed. The reason being that Wildlife Services work to protect the interests of animal farmers, or as they call it 'agricultural output'. Bears also don't pay for hunting licences.

So what if we did the opposite and actually protected wild predators?

In the 1920s, wolves became locally extinct in Yellowstone National Park due to hunting and ranchers in the area, which in turn led to the Yellowstone elk population growing. Efforts to curb their numbers through hunting, even though it had actually been outlawed in the national park, proved to be ineffective and unpopular, and ultimately failed.

Then, in 1995, wolves were reintroduced into Yellowstone, with their numbers reaching a natural plateau over time. Importantly, the population of elk also began to reduce and has since stabilised. This has led to degraded habitats recovering and regenerating, riverbeds becoming healthier, and populations of animals such as beavers, birds and bison increasing as well. There has also been an increase in the number of herbivores of other species thanks to less competition for food once the elk numbers came down, and, fascinatingly, the introduction of wolves reduced the number of coyotes, which in turn has led to there being more fawns, rabbits, mice and other small rodents, as they are not being hunted and killed by coyotes in the same way they were before.

Whereas hunting for population control is a reactive measure, as it means killing animals who are alive to try and bring down their numbers, reintroduction is a proactive measure, as it deals with the underlying cause and allows species numbers to naturally balance to match the limitations created by the environment.

Interestingly, one of the arguments against a reintroduction scheme is related to the impact it has on the prey animals. Obviously being hunted by a wolf is not pleasant for an elk, but neither is starvation. Nor is it nice for a fawn or a rabbit to be hunted by a coyote, which now happens less as a result of the wolves being reintroduced. I'm all for what we do in nature being analysed from the perspective of what it means for the suffering of individual animals in these systems. After all, environmental pursuits shouldn't be undertaken with the view that animals are simply metrics or cogs in our ecosystems, and their suffering and lives should also be factored into any environmental measures that are undertaken. However, the complexity of the natural world means that quantifying net suffering is challenging, and it is hard to say if the reintroduction of natural predators definitely increases suffering overall.

Perhaps unsurprisingly, the two groups most opposed to the reintroduction in Yellowstone were hunters and ranchers. But hunters are all about conservation and balancing population sizes, right? If hunters care about sustainability and protecting the natural world, why have they historically been the ones opposed to measures that would provide the most benefit to the environment?

People don't really hunt because they're worried that deer or elk are going to eat too much vegetation and stop songbirds from having a habitat. If hunters were genuinely concerned about the issues they say they are, such as conservation, animal suffering and sustainable food, they would eat only plants in addition to the animals they hunt, and they would pull the trigger with regret and remorse, not glee. As I said before, people hunt because they enjoy it and want to eat animals, which means the pretence that hunting is done with the best interests of animals at heart is exactly that – a pretence.

'HUNTING IS MORE ETHICAL THAN VEGANISM'

Even if the sustainability arguments don't hold up, what about the ethical ones? After all, a hunter only kills one animal, so don't vegans kill more animals as a result of the farming process?

When we addressed the argument that vegans kill animals for crops (see 'The "What-Abouter"'), one of the data points we looked at was that one hectare of land can produce 1,000kg of plant protein (although some yields are even higher) and would result in 15 animals being killed. Even though I disputed the methodology used to come up with this number of animal deaths, if we use it for the purposes of this debate, the ethical argument for hunting is immediately redundant. An average-sized white-tailed deer provides around 23.67kg of meat.[5] Deer flesh is approximately 24 per cent protein, meaning you get around 5.7kg of protein from an average white-tailed deer. So to get 1,000kg would mean having to kill 176 deer. This would make hunting responsible for nearly 12 times as many animal deaths per 1,000kg of protein as the contentious crop-death numbers. In reality, that difference is most likely going to be even larger.

So not only is hunting not a sustainable alternative to veganism but it also results in far more animals being killed.

'FARMED ANIMALS WOULD EAT US IF THEY WERE ABLE TO'

First of all, most of the animals we eat are herbivores, so this argument is utterly absurd. If you lie in a field with some sheep, they're not going to consider you a buffet. One of the main reasons we farm the animals we do is precisely because they pose a smaller risk to us. Herbivores are much safer and easier to subjugate than a predator who would potentially kill and eat us.

Ironically, it is the animals who actually do from time to time kill humans that we often feel the most outrage over the deaths of. Cecil the lion, who was hunted by a dentist called

Walter Palmer, is a good example. There was global condemnation around Cecil's death, but a lion poses a vastly higher threat to a human than a lamb does. Sharks can also kill humans – there were eleven shark-related fatalities in 2021, for example. However, many of us view shark-fin soup as a deeply immoral food to consume.

We don't view the worth of life of these animals based on the threat that they pose to us. We view these animals as possessing enough moral worth that killing them becomes an act deserving of condemnation. This is because we also recognise that we should hold ourselves to a different moral standard to that of other animals. The problem with farmed animals is that we don't grant them the same level of moral worth. While the life of one lion should be protected, farmed animals can be killed by the tens of billions every year.

The bottom line is, we kill far more animals than animals kill us. In fact, a conservative estimate is that we kill 2.5 billion animals for food every single day,[6] which means if we were killed at the same rate we kill animals, humanity would be wiped out within four days. The fundamental question we should be asking is not 'Would sheep eat us if they were able to?' but 'Why do we eat them when we don't have to?'

'VEGANISM IS ONLY POSSIBLE THANKS TO MODERN TECHNOLOGY'

While there have been vegans throughout history, it could be argued that with supplements, fortification and a wider variety of plant-foods available now, the ability to live an optimal plant-based diet is something that has only come about because of advancements made in our modern world. But even if that is the case, what does it really matter?

Modern technology is responsible for advanced health-care, global communication and large-scale infrastructure. Even outside of these life-changing innovations, it is also responsible for forms of entertainment, music, cinema and other enriching things that make life better and more enjoyable as a result.

I always find myself scratching my head when I hear about veganism only being possible in the modern world from people on podcasts that are only able to exist because of advancements in technology. You'll have to excuse my cynicism if I don't find this argument at all convincing, especially when it tends to be made by people who have popped up on social media to promote the idea that even looking at a vegetable will cause your muscles to disintegrate.

Even if veganism is only possible because of modern technology and knowledge, that does absolutely nothing to discredit it. Our understanding of vitamins, minerals and nutrition in general only exists because of modern knowledge and technology. The point is we can be healthy vegans, and it is the moral and right thing to do.

THE HISTORIAN

'WE'VE EATEN MEAT FOR THOUSANDS OF YEARS'

We've done lots of things for thousands of years, including killing and raping one another. Are those things also morally permissible because we have done them for a long time? Fundamentally, the question is, does the longevity of an action determine its morality?

Ironically, we often look back on the things that we used to do with disdain, disappointment and disbelief, even if those are things we did for a long time. So even if we approach this argument not from the position of it simply being a moral justification, but instead an appeal to tradition, or even a predetermined behaviour, the fact that we have not only changed so much as a species but also believe those changes to be preferable shows that we can and should modernise and alter the way we live for the better. The reason why we don't live in the way that we did thousands of years ago is precisely because we recognise that the longevity of an action doesn't make it moral or preferable as a consequence.

We've never lived optimally – we've simply lived as conveniently and opportunistically as we could at any given moment in history. Meat of course has played a part in that, but that doesn't mean it needs to any more. Throughout our history we have lived through times of food scarcity, which often meant consuming animal products. Nowadays, we not only have an abundance of plant-foods available to us, but we also have the nutritional and medical knowledge that allows us to be healthy and vegan at the same time (see 'The Amateur Nutritionist' and 'The Naturalist' for more).

'WE WOULDN'T BE ALIVE TODAY IF OUR DISTANT ANCESTORS HADN'T EATEN MEAT'

This could well be true. After all, meat helped us to survive during times when other food sources were scarce and we lived opportunistically, consuming the most convenient foods available to us. But are our ancestors good role models for how we should live in the twenty-first century?

We wouldn't be around today if our ancestors hadn't done lots of other things as well – murder, for example. Our ancestors killed one another, and they also killed earlier human species, including the Neanderthals. Does that mean that murder and even genocide are acceptable?

It doesn't take much to see that just because our ancestors did something doesn't mean that we should. As I've mentioned elsewhere, our ancestors often provide a good template for what we shouldn't do today. And even if we remove the moral component and instead view this argument as being more indicative of a predetermined behaviour, it doesn't hold up. Just as we have turned our back on many things our ancestors did, no longer consuming animals, especially when looking at the environmental impact, is of great benefit to our species. It doesn't need to be a part of who we are any more than living in caves does. The modernisation of our world has brought great benefits to our species, and if we are ever to become someone else's ancestors in the future, eliminating the farming of animals will need to be an integral part of our immediate future.

In other words, morally or simply behaviourally speaking, the consumption of meat was very different for our ancestors than it is for us. While they did it to ensure they could survive, we don't have that excuse and future generations will be grateful if we continue to modernise and progress in such a way that we stop using animals as food.

'EATING MEAT WAS THE REASON
OUR BRAINS DEVELOPED'

We sometimes like to believe our distant ancestors were ferocious meat eaters who gorged themselves on animal flesh, but in reality, when we were scavenging to survive, animals only made up a portion of our diet. The amount of meat that we were able to eat varied significantly, and we don't know for certain what caused our brains to grow and develop. While eating meat is one theory, it's far from being the only one.

One development that did take place around the time our brains began to evolve was the use of fire. This was a game-changer for our early ancestors for a number of reasons, one of the most important being that it allowed us to cook food, killing bacteria and making it safer to eat. It also widened our choices by making previously inedible foods palatable and in some cases more nutritious. This was particularly true of starchy foods like potatoes, wild grains and tubers, which are among the most carbohydrate-dense foods on the planet. Our body's main source of energy is glucose, a form of carbohydrate, with our brains using around 20–25 per cent of all our energy. Suddenly having access to a selection of foods that could provide our bodies and brains with a significant amount of its preferred energy source makes a compelling argument for why our brains developed and grew larger.

Meat might also have played a part in human development, and fire would have certainly made it safer to eat and easier to digest. Does it matter, though? Meat not playing a part in our brain development wouldn't in and of itself be a reason to forgo meat now. Likewise, if it did play a significant role, that wouldn't be a reason to consume meat today. Eating meat in prehistoric times was a matter of necessity; eating animal products in the twenty-first century is a choice.

And whatever the reason our brains developed, we could hardly make the argument that consuming meat is making us more intelligent today. So instead of using something that happened hundreds of thousands of years ago as a justification for eating animal products, how about we deal with what is tangible and happening right here in the present?

'NO CIVILISATION IN HISTORY HAS BEEN SOLELY PLANT-BASED'

This does certainly seem to be true. However, there were vegetarian cultures in the past, such as those found in ancient India, and there were prominent vegetarian groups and movements, such as in ancient Greece. In fact, the phrase 'Pythagorean diet' was once used to describe vegetarians due to the fact that Pythagoras didn't eat animals and advocated for others to abstain from doing so as well.

Fascinatingly, the consumption of many animal products was actually a huge taboo in Japan for around 1,200 years, only becoming socially accepted again in the nineteenth century. This was because of religious and practical reasons, and the fear that consuming certain animals was unclean. Plus, there might well have been cultures throughout history that ate only plants, but it is difficult to determine one way or another due to a lack of evidence.

All that being said, it is true that there have not been any solely vegan civilisations or cultures that we know of. But there weren't any previous civilisations that had the medicines we have, or the technology, or the infrastructure either. All of these are examples of things that have contributed positively and are unique to the times we live in. In other words, life today is improved as a consequence of being able to do things that previous civilisations could not.

I'm not sure how many people who use this argument would refuse a life-saving procedure on the grounds that it wasn't possible in days gone by. 'Sorry, Doctor, if the ancient Greeks couldn't perform life-saving keyhole surgery, then I'm afraid I just won't allow it.'

So while abstaining from all animal products, especially on non-religious grounds, might be something more commonly associated with our modern world, what exactly does that prove? Even if we take the position that it wasn't possible for previous civilisations and cultures to be vegan, all that demonstrates is that we are able to make different choices to people who came before us. If anything, that sounds like a positive thing to me.

'THE BIBLE SAYS I CAN EAT ANIMAL PRODUCTS'

One of the most astonishing stories I have heard so far about eating meat and religion comes from Catholicism. In the Catholic calendar there are certain dates where it is forbidden to eat meat, with the exception of fish, along with some other less well-known exceptions.

In the eighteenth century, capybaras, which are the world's largest rodents, were a popular food source in some areas of the world. Because eating meat was forbidden during Lent, some clergyman wrote to the Vatican asking that capybaras be reclassified as fish. Astonishingly, the Vatican approved. Beavers were also considered to be fish for the same reason. This is because these species, while clearly not fish, do spend time in water.

How much time one must be in water to be reclassified as a fish has not yet been defined by the Vatican. Considering that Catholics believe God literally created these animals, it was certainly audacious to so flagrantly bend the rules in front of their maker. It also made a complete mockery of their claim to be abstaining from meat.

This all speaks volumes about our desire to eat flesh. People will run the risk of encouraging God's wrath if it means for a small period of time they can continue eating meat.

More generally, the notion of animal rights is a concept that often draws the ire of certain religious adherents; for example, the homophobic evangelical pastor Steven Anderson:

> But I'll tell you what this veganism is, it's a substitute for biblical morality . . . If somebody wants to be vegetarian or vegan for whatever reason, OK, great. But if you're going to say it's because God wants us to, that's a doctrine of devils . . . So if a vegetarian starts commanding me to abstain from meat, that's when I'm going to say, 'Hey, get thee behind me, Satan.'

Although I'm not in the business of commanding anyone, I am obviously an opinionated advocate for veganism. I wonder if that makes me a disseminator of the devil's doctrine? Actually, I have been accused of being the devil. While engaging in a debate with a sign-wielding religious protestor in the USA, I was accused of being Satan in human form come to challenge his faith. This was because I pushed back when he said that the reason God created crocodiles was so they could be turned into bags. Needless to say, the conversation didn't end especially well.

I am however keen to not straw man all religious adherents by claiming they all believe such ridiculous things. So let's take a broader look at the relationship between what we do to animals and Christianity.

Even if we overlook animal welfare, can God be happy with the destruction of the planet He created? As animal farming is the number-one cause of deforestation, habitat loss and species extinction, the idea that God condones the practice seems contradictory. After all, 'God saw all that He had made, and it was very good.' The environmental arguments against animal farming on their own therefore speak to the virtues of those who are religious becoming vegan.

Some Christians like to argue that God gave us dominion over animals and the Earth. But dominion doesn't provide a justification to tyrannise, subjugate or destroy. Parents have dominion over their children, but that doesn't make parental child abuse acceptable. Dominion comes with responsibility. It comes with power, but to wield power in an authoritarian way is to act immorally and irresponsibly. Rather than dominion being a justification for violent supremacy, it should be viewed as the responsibility to steward and act with integrity and accountability.

Christians also argue that only man is made in the image of God, but that doesn't mean we have the right to harm non-human animals. After all, God gave them the ability to feel pain and to suffer. Why would a benevolent God accept us causing suffering and pain to others, even if they are not human?

And if God is happy with us causing suffering to animals, is there a line or a limit? Is there any violence towards animals that God does not find permissible? If yes, why does He not think forcing them into gas chambers, cutting their throats and mutilating them is also wrong? After all, these are among the most violent things we can do to an animal.

Farms and slaughterhouses are places that are closer to being literal depictions of hell than they are heaven. Violence, bloodshed, fear, pain, anxiety, terror, torture, slaughter – what is godly, spiritual or righteous about any of that? As the Bible states, 'The Lord is good to all, and His mercy is over all that He has made.' What mercy is found in a slaughterhouse?

Let's also not forget that the animals we farm are selectively bred. If God is perfect and doesn't make mistakes, is it not blasphemous that we have hijacked the reproductive systems of His creatures and then genetically altered them so that they are no longer as He created them? Is this not a criticism of God, calling His work imperfect, as we have changed these animals to better suit our purposes? To me, this seems to be the definition of blasphemy – it is showing utter disrespect to God, who doesn't make mistakes. Except He made the mistake of creating chickens that don't lay enough eggs or who don't grow fast enough. He made the mistake of creating cows who only produce enough milk to feed their own children. And He made the mistake of creating sheep who shed their wool. He also didn't make pigs, turkeys, ducks, dogs, cats and other animal species perfectly either, hence why we have selectively bred them all. Even with His divine omniscience, God failed to create the 'right' animals.

There are, of course, historical precedents in the Bible for eating animal products, and while I might argue that eating animal products is immoral now, when Jesus was purported to be alive and the Bible was written were very different times, and the moral standards of today are different as well. It is for this reason that claiming that eating animal products is immoral now is not casting judgement on the actions of

Jesus in the past. And although it is true that eating animal products has not always been immoral, I believe the morality of eating animal products should change as the world changes, especially when you consider that the behaviours and actions of Christians have changed in the past 2,000 years in relation to other matters.

While the Bible does contain passages that condone the consumption of animals, there are others that cast doubt on that assertion. Perhaps most strikingly, Adam and Eve didn't slaughter animals and lived peacefully among them in the Garden of Eden. Plus, the depiction of God's mountain refers to animals living peacefully alongside one another with the lion eating straw with the ox. It even states, 'They shall not hurt or destroy in all my holy mountain.'

Fundamentally, because eating animal products is one of the leading drivers of the most serious environmental problems we face, and considering it is the leading cause of human-inflicted animal suffering, harm and death, I would argue that what we eat today undermines the purported overarching principles of Christianity – benevolence, love, forgiveness, and so on.

Furthermore, there are vegans within all religions. There are Christian vegans, Muslim vegans, Hindu vegans, Buddhist vegans, Sikh vegans, Jewish vegans, and so on. Veganism is not a barrier to religious adherence, and even though this might not be the claim many religious adherents make, this understanding reinforces the fact that whether or not we eat animals is still a choice in front of us. You are not less of a Christian or less of a Sikh if you are vegan. There are no religious mandates to eat animals or their secretions, even if there are some passages from texts written thousands of years ago that permit it during that time. Veganism is certainly not an anti-religious position. If anything, it is simply an extension of the values that many religions preach.

If we can cause less harm to God's creatures today, why wouldn't we? If we can cause less harm to God's world, His perfect creation, then why wouldn't we? And even if some adherents would claim it is blasphemous or disrespectful to elevate non-human animals to a moral status equivalent to that of humans, doing so is not a requirement in order to be vegan. One can be vegan and still consider humanity to be made in God's image and above other animals.

Veganism is merely taking the position that animals are deserving of moral consideration in their own right, and their sentience (given to them by God) makes them individuals worthy of respect, dignity and liberty to not be exploited, subjugated, harmed and slaughtered by us.

Plus, if Catholics can eat rodents and pretend they're fish and that's OK with God, then vegans can just eat plants.

'HALAL AND KOSHER ARE ETHICAL FORMS OF SLAUGHTER'

These methods of slaughter are held up as being humane and ethical by religious adherents. Both methods involve cutting the throat of an animal without prior stunning. Interestingly, the act of cutting an animal's throat in the manner set out by these teachings might have been the quickest and most painless method thousands of years ago when people relied on animals in their diets. After all, we didn't have any other means to kill animals, except for things like bludgeoning them or suffocating them, which would have been crueller. However, to say that halal or kosher slaughter methods are kind to animals in the modern world, when we don't have to kill and eat animals in the first place, is to ignore everything that has changed over the previous millennia.

In fact, the acknowledgement that we should seek to reduce animal suffering, as is the purported aim of such slaughter methods, should provide a compelling argument for Muslims and Jews to now become vegan, as the best way to reduce animal suffering is to not eat animal products. The same comment could be made to those who advocate for stun slaughtering as well. If we really want to stop animals from suffering, the simple solution is to not exploit and slaughter them in the first place.

Even religious festivals like Eid and Kaporos, which have historically revolved around the mass slaughter of animals on the street, are able to achieve their aims without the spilling of blood. There are many vegan Muslims and Jews who offer different forms of non-violent sacrifice while still adhering to the core principles behind these festivals.

THE AMATEUR NUTRITIONIST

'VEGANS LACK PROTEIN'

Protein. The nutrient of all nutrients. The building blocks of the body, and the building blocks of so many anti-vegan arguments. Lots of meat eaters think the topic of protein is to vegans what kryptonite is to Superman. But in reality it's simply not a concern at all. Here's why.

Protein is formed of amino acids, and of those our body needs to be healthy, nine are called essential amino acids. This means that they must be acquired through diet, as they are not created by our bodies. All plant-foods contain all of the essential amino acids that we need from our diet.[1] However, with some exceptions, most of these foods are not considered complete proteins, as they don't contain enough of all the essential amino acids. This means that from a protein perspective, we might not get enough of all the essential amino acids we need from just one plant-food.

Luckily for us, we don't eat only one food. We eat a variety of different foods, meaning that as long as we get our recommended caloric needs and do so by eating a range of different foods, with some good plant-based protein sources such as beans, lentils, chickpeas, tofu, tempeh, seitan, nuts or nut butters, to name but a few, we can get all the protein we need. The only way that protein deficiency would become an issue is if we ate significantly less than we needed to be healthy or if we just ate a couple of types of foods. In other words, don't just eat blueberries and then wonder why you're not getting enough protein.

To get the protein that we need to be healthy, and even the quantity we need to build muscle and be athletic, is absolutely possible on a plant-based diet. The recommendations from health organisations and nutrition associations don't say anything different, hospitals are not filled with protein-deficient vegans, and the protein-powder market is not being supported by vegans who are on the floor desperately dragging themselves to health food shops with the last ounces of strength they have left.

The largest study comparing the nutrient intake of vegans and meat eaters showed that meat eaters got 75.8 grams of protein per day compared to vegans who got 72.3 grams, meaning there was little difference between total protein intake. And in the case of the meat eaters, almost 60 per cent of their protein intake actually came from plants.[2]

However, not content to simply concede that you can get protein from plants, some anti-vegans make the claim that plant protein is of a lower quality. As such, the protein argument is still used as an excuse to kill animals. But is this true? Is plant protein actually inferior?

It's important to recognise that once in the bloodstream, the body has no idea what the source of the amino acids you have consumed was. It doesn't know if they came from tofu, lentils or animal flesh – it takes those amino acids and uses them all the same. In fact, a 2021 study showed that vegans and meat eaters eating 1.6 grams of protein per kilogram of body weight per day (for reference, around 0.8 grams per kilogram of body weight per day is the recommended minimum intake) found that there was no difference between the two groups with regard to muscle mass increase and muscle strength.[3]

As we pass 50 years old, it is advised that our protein

intake increases in order to help maintain muscle mass and bone density, which would mean adding in a couple more servings of protein-rich plant-foods, such as legumes, to our diets each day. A 2019 study looking at men and women aged 40 to 80 further found that it was the amount of animal or plant protein intake that affected muscle mass, not the type of protein.[4]

The main source of the argument that plant protein is inferior to meat protein stems from something called DIAAS (Digestible Indispensable Amino Acid Score), which attempts to determine the protein quality of different foods by looking at amino acid levels and digestibility. However, it has some glaring flaws that have led to calls for the way we view protein quality to be modernised.[5]

To begin with, the research that is used to define protein absorption for DIAAS comes from animal studies on rats and pigs, fast-growing animals whose protein requirements and usage differ from that of us humans. In a comprehensive overview of protein scoring, the authors of the review called for 'new data on protein material' due to the animal studies, with the authors pointing out that animal feed differs when compared to food products intended for humans.[6]

One such example is that these animal studies analysed the consumption of raw plant-foods. However, we don't eat raw legumes, such as beans, lentils and chickpeas, and we don't tend to eat raw grains either. Anybody's dinner consist of raw kidney beans with raw rice? The reason we cook these foods is to make them digestible, and the bioavailability and digestibility of the proteins in these plant-foods is increased as a result of soaking, cooking and processing them.

One of the other primary issues with DIAAS is that it overstates the protein content of some animal products[7] and understates the protein content of some plant products due to a flaw in its methodology.[8] DIAAS is also reductionist in the sense that it looks at foods in an isolated manner, when of course we eat a variety of different foods that complement one another. For example, porridge made with soya milk is classified as 'high-quality protein' even though DIAAS understates the protein content of soybeans.

When it comes down to it, if there were a significant difference in the bioavailability and quality of plant protein over animal protein, the nutritional guidelines would reflect that fact. And rather than plant protein being an issue, it's actually the animal protein we consume that is causing us problems, with the scientific literature showing us that we need to consume less of it.[9]

So, no, animals don't need to be slaughtered for us to get protein, and the credible science is clear and unambiguous on this.

'VEGAN DIETS DON'T CONTAIN A NATURAL SOURCE OF VITAMIN B12'

In the UK, due to the success of Veganuary, the Agriculture and Horticulture Development Board (AHDB) now run a campaign in January called 'We eat balanced', which, amongst the rest of its pro-meat rhetoric, homes in on natural B12. For example, in one of their adverts they state, 'Did you know that beef, pork, lamb and dairy are natural sources of vitamin B12, an essential vitamin not naturally present in a vegan diet?'

So is this a good reason to not be vegan?

Humans can actually produce their own B12; however, it doesn't get absorbed by our bodies and is instead excreted out. This means that there is in fact a natural source of B12, but it would mean practising coprophagy (otherwise known as eating your own faeces). I know that some people say vegan food tastes bad, but I promise it's not that bad. So, in the absence of a natural B12 source, vegans get their B12 from fortified plant-foods like plant milks and from supplements.

The thing is, the vast majority of the animals we farm are also mono-gastric like us, meaning they need a dietary source of B12, and because they're intensively farmed, the B12 they get also comes from fortified foods and supplements. Ruminant animals like cattle and sheep are able to produce B12 themselves, but in order to do so they need to have the mineral cobalt in their diet. In theory, cobalt is found in soils, but in actuality this isn't necessarily a given because soil is often cobalt-deficient. As a result, many pastures have cobalt added to them. Alternatively, ruminant animals are given the cobalt or B12 directly. In other words, unnatural B12 is a ubiquitous reality of our food system.

So the AHDB's recommendation that we get natural B12 from meat is disingenuous. By emphasising the consumption of natural B12, the implication is that natural is somehow better. Obviously this is not the case. As I mentioned earlier, faeces is a natural source of B12, but it would obviously not be a good idea to eat it. Just to be safe though, with their insistence on natural B12, it's probably not advisable to accept a dinner invitation from the heads of the AHDB. It also makes zero sense to consume B12 filtered through the body of an animal when we could just take the supplements ourselves. Even the AHDB list supplements and fortified foods as sources of vitamin B12 on their website.[10]

So, while vitamin B12 might be portrayed as the final nail in the malnourished vegan's coffin, this argument is simply a load of natural-B12-containing faeces.

'VEGANS LACK IRON AND HAVE HIGHER RATES OF ANAEMIA'

I ron is an essential mineral that helps transport oxygen to our bodies' cells and is incredibly important when it comes to being healthy. Among other things, iron deficiency can cause anaemia, fatigue, impaired brain function and fainting. As a result, it's very important that we get enough iron from our diet, especially pre-menopausal women. In fact, while men and non-menstruating women need 8.7 milligrams of iron a day, pre-menopausal women require 14.8 milligrams.[11] The good news is that there is no reason why vegans can't meet these requirements.

There are two types of dietary iron: haem and non-haem. The former comes from animals and refers to haemoglobin, and the latter comes from animals and plants. Haem iron is more easily absorbed by the body but comes with health risks. It has been associated with colorectal cancer[12] and heart disease.[13] Iron is also a pro-oxidant, which means excess amounts can cause damage to cells and organs.

Good sources of non-haem iron include wholegrains, legumes, dark leafy greens, seeds and dried fruits. An easy way to boost the absorption of non-haem iron is to combine it with a good source of vitamin C. This could include squeezing some lemon juice on your food or adding some berries to your breakfast. Vegetables like kale and broccoli are also good sources of vitamin C, as are fruits like bell peppers, so having a serving of these with your dinner can help as well. Drinks like coffee and tea, on the other hand, have been shown to inhibit iron absorption, so it's advisable to not drink them during meals.

Some studies have shown a slightly increased risk of

anaemia in people with low or no red meat intake,[14] while others have found no difference between vegans, vegetarians and meat eaters when it comes to iron deficiency.[15] Interestingly, an analysis of 24 studies showed that non-meat eaters are more likely to have lower iron stores. However, the study concluded that due to high iron stores being a risk factor for diseases such as type 2 diabetes, this could actually be positive. In their own words, 'non-vegetarians should regularly control their iron status and improve their diet regarding the content and bioavailability of iron by consuming more plants and less meat'.[16]

It's important to note that iron deficiency is one of the most prevalent nutrient deficiencies globally, affecting people in both low- and high-income nations. And we should acknowledge that it is entirely possible to be iron-deficient as a vegan, although it is entirely possible to be iron-deficient as a meat eater as well. Because of this, we should all make sure that we are including good sources of iron in our diets every day, especially if you are pregnant or, as I mentioned at the start, a woman who menstruates. However, having concerns about iron deficiency definitely does not mean that we should all eat meat.

'VEGANS LACK CALCIUM AND HAVE HIGHER RATES OF BONE FRACTURES'

Because calcium has become synonymous with dairy, it is perhaps understandable that one of the biggest fears around eliminating all dairy products would be a lack of calcium. This has led to articles and dairy-industry press releases warning against plant-based diets. However, is calcium deficiency actually a real concern for vegans?

In the UK, it is recommended that we get around 700mg of calcium a day. Fortified soya milk will provide around 120mg of calcium per 100ml. So just one 250ml glass of soya milk provides more than 40 per cent of the recommended calcium intake. Alternatively, just 100g of tofu that is made with calcium can contain 350mg of calcium or 50 per cent of the daily intake. Other good sources of calcium include other legumes; almonds and Brazil nuts; seeds like chia and flax; tahini; calcium-fortified vegan yoghurts; dried fruits like figs; vegetables like kale, bok choy and broccoli; fruits like oranges and tangerines; oats; and calcium-fortified foods like bread and cereals.

But wait, what about bone fractures?

The issue of vegans and bone fractures has become a recurring talking point, mainly because of a 2020 study that showed an association between veganism and increased bone fracture risk.[17]

One of the biggest variables that determines bone density and health is actually weight. In fact, one of the best ways to build bone health is to do weight-bearing exercises. A study looking at the relationship between body weight and bone mineral density (BMD) concluded: 'Adults with obesity had significantly higher BMD than healthy-weight subjects.'[18]

Although the 2020 study showed an increased risk of bone fractures in vegans, the researchers noted that when adjusted for BMI the associations became weaker. Plus, the researchers were unable to make a comparison between vegans with high BMIs and meat eaters with low BMIs because there weren't enough overweight vegans.

On top of this, the data was actually derived from a cohort study that started back in the 1990s, when fortified plant-foods were far less common and awareness about how to be a healthy vegan was also lower than it is today. This is a potential reason why the vegans in the cohort had significant rates of B12 deficiency, which as well as potentially being important for the findings of the study also points to a wider lack of nutritional education in general.

Another important thing to note is that calcium on its own won't safeguard us against osteoporosis or weak bones. Vitamin D is also vital when it comes to bone health. It is advised that vegans and non-vegans alike, especially in countries like the UK where there is less sun, take a vitamin D supplement. A lack of vitamin D could also be a factor for the higher rates found in the 2020 study due to the previously mentioned lack of fortified plant-foods. Unfortunately, the study didn't look at vitamin D. Vitamin B12 and protein are also essential for good bone health, as is exercise, including both weight training and impact exercise like jogging and skipping.

Interestingly, data from the same cohort study has also shown that the vegans had a lower risk of heart disease, diabetes, cancer, diverticular disease and cataracts. They also had lower blood pressure, lower LDL-cholesterol and a healthier ratio of saturated fat intake.[19]

So if the vegans in this study were not the healthiest vegans, with low levels of some essential nutrients and even low fibre intake compared to what would be expected for those eating a healthy plant-based diet, what happens when health-conscious vegans are analysed?

A paper released in 2021 found no increase in hip fracture risk for vegan men compared to non-vegan men. However, an increased risk for vegan women was reported until calcium and vitamin D were factored in. When vegan women were taking supplements and getting adequate amounts of calcium and vitamin D, the risk completely disappeared for vegan women as well.[20] So the best way to reduce the risk of bone fractures is to simply make sure you are getting enough calcium, vitamin D, vitamin B12 and protein, incorporate some regular weight-bearing exercises into your routine, and make sure to maintain a healthy weight.

The idea that cows' milk is essential for healthy bones is an interesting one when you consider that we've only consumed dairy for around 10,000 years, which is a fraction of the time that humans have existed, plus almost 70 per cent of humanity is lactose intolerant.[21] It would also be a pretty terrible design flaw if bone health was predicated on drinking the milk of an entirely different species of animal who only produces milk to feed their own offspring. The argument seems to be that we have to engage in a kind of interspecies breastfeeding to be healthy. It would be pretty weird to see someone lying underneath a cow in a field suckling directly from their udders, but what's the actual difference between doing that and buying milk in a carton from a supermarket? The alternative is eating calcium-rich plant-foods, drinking plant milks that contain calcium, and making sure we are conscious of our bone health. It's your call which option makes most sense.

'VEGANS LACK OTHER ESSENTIAL NUTRIENTS'

Even though I've covered the main nutrients that are often perceived as being deficient in a plant-based diet, I can already hear people saying, 'But what about omega-3?' or, 'But I saw a video that says vegans are deficient in 50 nutrients and we should all be eating raw liver?'

When it comes to omega-3, you can get ALA, a short-chain omega-3 fat, from plant-foods such as flaxseed, chia seeds, hemp seeds and walnuts. The body can convert ALA into EPA and DHA, which are long-chain omega-3 fats. However, the rates of conversion can vary, so it's advisable to take an algae-oil supplement to make sure you are getting a good amount of

EPA and DHA. Interestingly, fish do not produce these long-chain omega-3 fats themselves either, so even if you consume fish, these nutrients originate from algae anyway.

So, in essence, we can leave the fish alone, protect our oceans, and reduce antibiotic use and the environmental problems of fish farming and wild fishing by just taking an algae-oil supplement instead of eating fish or taking a fish-oil supplement. This provides an easier and more convenient way of getting these nutrients and ensures that we are getting the amount we need, especially as the levels in fish can vary and most people don't eat fish every day.

Now to the wider point about the healthfulness of plant-based diets in general. There can be a lot of noise about health online. Sometimes it can feel overwhelming to be told one thing and then seeing someone online telling us the complete opposite. However, there's a reason why plant-based diets are endorsed as being healthy by the leading health organisations around the world, including the NHS,[22] British Dietetic Association,[23] the Academy of Nutrition and Dietetics,[24] Harvard Medical School,[25] Dieticians of Canada[26] and the World Health Organization[27], many of which emphasise that plant-based diets can actually improve health outcomes and are suitable at all stages of life.

Ultimately, when the leading authorities on a subject are telling us the same thing, it's a good indicator that there's some truth to what they're saying. Importantly, these determinations are based on a large body of evidence. This is what the collective field of research is showing, rather than an isolated and cherry-picked study. And assessing what the whole body of evidence is showing is the best way to understand what the recurring patterns are and to see what is actually true.

So, call me old-fashioned, but I think it's better for us to get our information from experts and not random people on the internet.

'VEGAN DIETS CONTAIN ANTI-NUTRIENTS SUCH AS LECTINS, OXALATES AND PHYTATES'

The conversation around these so-called 'anti-nutrients' is the epitome of viewing food in a reductive manner, because it neglects so many important aspects of the evidence and does not consider foodstuffs as a whole. An item of food contains many nutrients, and to claim that something is inherently bad because it includes a so-called anti-nutrient is like saying (to reference the point from earlier) that dog faeces is inherently good for you because it contains vitamin B12.

So, let's start with lectins.

Lectins

The fear about lectins, which are commonly found in legumes and wholegrains, is that they prevent the absorption of nutrients and can cause conditions like leaky gut. These concerns have been popularised by Steven Gundry, a physician and author of the book *The Plant Paradox*. The problem with his logic is two-fold. First, the science supporting a lectin-free diet is of exceedingly poor quality. Some of the data comes from studies in which the animals were given incredibly high quantities of lectins, more than you would get from a normal diet, which caused them health problems. Force-feeding rats excessively high quantities of something is not a good basis for claiming that thing is inherently bad for humans. A lot of the other data comes from Gundry and his patients. However, this work is not peer-reviewed and published, there is no reference to control groups, and (at the time of writing) the research linked on his website contains no methodology, only abstracts. All of these factors render the research more or less useless.

Second, lectin levels are dramatically reduced in foods when they are soaked and/or cooked. In fact, a study looking at the rates of lectins in dried beans, the foods often touted as containing the most lectins, found that the 'cooking of beans completely destroyed the lectin activity'.[28] And other research has indicated that lectins could actually play a role in cancer prevention,[29] protecting human cells from damage caused by free radicals[30] and gastrointestinal metabolism.[31]

If lectins were so harmful, there would be clear evidence showing the consumption of legumes and wholegrains as having a negative impact on people's health. However, the opposite is the case. Legumes have been shown to be 'the most important dietary predictor of survival in older people of different ethnicities'.[32]

Plus, a comprehensive analysis of 66 studies showed that those consuming three to five servings of wholegrains per day had a 26 per cent lower risk of type-2 diabetes and a 21 per cent reduction in heart disease risk when compared to those who rarely or never consume wholegrains.[33]

As one past president of the American Heart Association put it, leaving out plant-foods because of lectins goes 'against every dietary recommendation represented by the American Cancer Society, American Heart Association, American Diabetes Association and so on'.[34]

But hey, if you're really worried about lectins, Dr Gundry has you covered. For only $80 a month you can purchase his 'Lectin Shield' supplements . . .[35]

Oxalates

Oxalates can reduce calcium absorption, and oxalate stones can form when calcium and oxalate bind and form crystals. In fact, the most common cause of kidney stones is calcium oxalate. Because of this, people who are prone to kidney stones are sometimes recommended to go on a low-oxalate diet. That being said, studies have also shown that dietary oxalate and kidney stones were found to only have a moderate association, with the authors concluding that dietary oxalate is not a major risk factor.[36]

The highest oxalate content is found in foods such as spinach, Swiss chard, beet greens and almonds. However, for the average person oxalates are not a reason to avoid these foods altogether, especially as the levels in these products can be reduced very easily. For example, cooking spinach and Swiss chard causes up to an 87 and 85 per cent reduction in oxalate levels respectively.[37]

Almond and cashew milk have been shown to have higher kidney-stone risk factors than other plant-milks, with oat, macadamia and soya milk comparing with dairy milk in terms of their overall risks.[38]

In practical terms, what does this mean? The simplest thing is to simply diversify the plant-foods you consume. Rather than only drinking almond milk, mix it up and have oat milk and soya milk as well. When it comes to vegetables, don't just eat raw spinach and beet greens every day. Instead, sometimes cook them when you do consume them and at other times opt for other greens, such as kale, bok choy and broccoli. Plus, dehydration is a contributing factor to the emergence of kidney stones, meaning drinking plenty of water is one of the best safeguards we have against their formation.

It's also worth noting that red meat, organ meats and shellfish also have high concentrations of a compound called purine that can lead to higher production of uric acid, which can then cause uric acid stones, another type of kidney stone.

Phytates

Phytate, often referred to as phytic acid, is commonly found in foods such as wholegrains, nuts, seeds and legumes, and it is sometimes argued that it can decrease the absorption of nutrients such as zinc, calcium and iron. However, it's not quite as simple as that.

The phytic acid content of foods can be reduced when they are soaked, cooked, fermented, pickled and sprouted. For example, cooking legumes can reduce their phytic acid content by up to 80 per cent.[39] Considering that legumes include foods such as lentils, beans and peas, the phytic acid content of them in their raw form is not representative of what eating them will actually mean for our ingestion of phytic acid. Plus, a diet high in phytic acid has been found to reduce the negative effect of phytate on non-haem iron absorption.[40] In other words, the body appears to adjust and actually increase non-haem iron absorption to compensate for a higher phytic acid intake. Furthermore, as mentioned previously, consuming vitamin C actually boosts the amount of iron absorption in our body, meaning that although phytic acid can have a negative impact on vitamin and mineral absorption, there are ways to boost absorption as well.

However, the main reason why this conversation isn't so simple as the 'phytic acid is bad' rhetoric suggests is because phytates can also have positive health effects. Phytic acid is an antioxidant, meaning it can help to remove free radicals that can contribute to the development of cancer and other diseases. This is one of the reasons why it is thought that phytates could actually reduce the risk of colon cancer.[41] Phytate-rich foods have also been shown to decrease osteoporosis[42] and kidney-stone risk,[43] and phytate could also be helpful in the treatment of type 2 diabetes.[44] Overall, "the benefits of phytate-containing foods to human health exceed the impacts on mineral absorption."[45]

'PLANTS ARE HARMFUL TO OUR HEALTH AND CAN KILL US'

Beyond lectins, oxalates and phytates, the conversation around plants goes even further, because for many pro-meat advocates, the problem is not just that plants contain 'anti-nutrients' – it's that 'plants are trying to kill you'.[46]

For example, did you know that there is a molecule called dihydrogen monoxide that vegans consume every day and is used in the production of plant-based alternatives? However, it's also used as an industrial solvent and in nuclear power plants. Not only that, but it can cause suffocation, accelerates corrosion and rust, and is in the bodies of everyone who has cancer. I think we'd all agree that this is a molecule we should definitely avoid. Except dihydrogen monoxide is simply a chemical name for water. This aptly demonstrates that if you give something an unfamiliar name and then portray it in a negative way, it is easy to make it sound scary. It's a good example of how we can be manipulated into fearing something unnecessarily.

This is something that anti-vegans and carnivore advocates do all the time. For example, some people claim that cruciferous vegetables like Brussels sprouts and broccoli contain dozens of carcinogens. Yet a study published in the *Annals of Oncology* showed that the consumption of cruciferous vegetables was associated with a lower risk of cancer.[47]

One of the main factors behind why cruciferous vegetables have been shown to be associated with lower cancer risk is because of a molecule called sulforaphane. Ironically, carnivore advocates claim that sulforaphane is bad for us as well, blaming it for a whole of host of things. Yet sulforaphane has been extensively studied due its benefits, including its anti-cancer properties.[48] Cruciferous vegetables also contain antioxidants that are antibacterial, antiviral and anti-inflammatory, and consuming cruciferous vegetables has been shown to reduce the risk of cardiovascular disease.[49] To put it simply, if cruciferous vegetables are trying to kill us, they're doing an exceptionally bad job.

Some anti-vegans also reference glycoalkaloids, naturally occurring compounds that are found in foods such as potatoes, tomatoes and aubergines and which are toxic to humans at high levels. The most well-known glycoalkaloid is solanine, which can be found in potatoes. However, fatal cases of solanine poisoning are very rare, especially as commercial varieties of potatoes are screened for solanine and our bodies are able to break down the chemical in cooked potatoes. Higher levels can be found in greening potatoes, but this simply means avoiding them and removing any potato sprouts.

A similar principle could be applied to apple seeds, as they contain cyanide, but to consume a lethal dose would mean consuming a huge number of them. Despite the risk being so extraordinarily low, creating a narrative around the dangers of apples by focusing on this one aspect and then failing to give it the appropriate context it needs could understandably cause confusion and doubt around the perceived safety of apple consumption.

Anti-vegans also frequently talk about the supposed dangers of phytoalexins, which are molecules released by plants as a defence mechanism. These people suggest that because one of the roles of phytoalexins is to prevent the plants from being eaten, this therefore means they are toxic for us to eat. However, one of the most well-known phyto-alexins is resveratrol, which has anti-carcinogenic, antiviral, neuroprotective, anti-inflammatory and antioxidant proper-ties.[50] Plus, a systematic review of phytoalexins and their pharmacological potential concluded that phytoalexins have 'tremendous potential in the treatment of various life-threatening diseases such as diabetes, cancer, brain damage and heart attack'.[51]

As you might have noticed, there's a pattern here.

More broadly than just isolating certain molecules within plants, carnivore advocates will claim that because there are plants in nature that can kill us or that should be avoided, this therefore proves that eating plants is not good for us. This is clearly a reach that ignores that two things can be true at once: there are plants that can kill us and there are plants that promote positive health outcomes for us.

Simply put, this argument rests on outlandish assump-tions and speculation. A claim has to be supported by data if it is to be taken seriously, but the outcomes shown by scien-tific studies looking at plant consumption don't show that they are trying to kill us – they show the opposite. A huge meta analysis of 76 studies that included more than 2.3 mil-lion participants showed that plant-based diets are beneficial for lowering the risks of major chronic diseases, including type 2 diabetes, cardiovascular diseases and cancer, as well as reducing premature deaths.[52]

If the plants we eat were toxic or trying to kill us, this would be reflected in the scientific literature. The fact that the opposite is true should be a very clear indicator of the ver-acity of such claims.

'SEED OILS ARE BAD FOR YOU'

Brandolini's law is something I've had to become very familiar with in my advocacy for veganism. It is often referred to as the bullshit asymmetry principle, because, as the law itself states: 'The amount of energy needed to refute bullshit is an order of magnitude bigger than that needed to produce it.'

There is no finer example of this than the conversation around seed oils.

In many ways it's not surprising that something like seed oils has for many people become the main reason for the world's health woes. After all, for those who advocate for eating meat, or simply want to continue consuming animal products, the narrative that meat isn't the problem and seed oils are is a very attractive one.

Anti-seed-oil advocates have claimed that 'vegetable oils are worse than smoking two packs of cigarettes a day',[53] and outlandishly argue that seed oils are responsible for causing sunburns and saturated fat from meat protects the body from UV rays. In fact, some carnivore influencers even claim that you should avoid sunscreen and that eating meat will protect you from any danger from the sun.

However, the most common claims about seed oils are that they are toxic, cause inflammation and lead to disease because they contain linoleic acid, an omega-6 fat that can be converted into another fat called arachidonic acid, which can in turn be converted into molecules that can play an inflammatory role. However, a systematic review of human studies showed that even when consumption of linoleic acid was increased by up to six-fold, no significant correlations with arachidonic acid levels were observed.[54] In other words, the conversion rate is very low in humans. And from the arachidonic acid that is converted, anti-inflammatory molecules are also converted, not just pro-inflammatory ones.

Anti-seed-oil advocates also point to research that was carried out on rodents, but mice and rats don't respond to linoleic acid the same way humans do, further undermining their argument. And they often share graphs showing that the prevalence of chronic disease has increased in step with population-wide seed-oil consumption. But you could also share a graph showing that the human population has increased alongside seed-oil consumption, but does that mean consuming seed oil increases fertility or sexual desire?

As always, we should ask ourselves: what does the balance of evidence really say about seed oil? A meta-analysis that looked at studies on linoleic acid and inflammatory markers concluded that 'virtually no evidence is available from randomized, controlled intervention studies among healthy, noninfant human beings to show that addition of linoleic acid to the diet increases the concentration of inflammatory markers'.[55] This is true even when looking at safflower oil, which has one of the highest percentages of linoleic acid in the seed oils consumed by people.[56]

But if it doesn't cause inflammation, what about chronic disease? A meta-analysis published in one of the world's top cardiology journals showed that people with higher levels of linoleic acid in their blood and fat stores have a lower risk of major cardiovascular events.[57] Systematic reviews and meta-analyses have also shown that linoleic acid is associated with a lower risk of type 2 diabetes,[58] high blood pressure,[59] cancer and mortality from all causes.[60]

And finally, what about if we compare polyunsaturated fatty acids with saturated fats? For example, what if people switched from animal-based saturated fat to canola oil (often referred to as rapeseed oil)? Canola oil is a particularly good comparison because it is commonly used in things like oat milk and plant-based alternatives. A study of over half a million people showed that substituting butter with canola oil was associated with lower cancer mortality, cardiovascular mortality and total mortality.[61] This is backed up by a meta-analysis of 42 clinical trials that showed that canola oil improved several cardiometabolic risk factors when compared to butter.[62] And if we expand beyond just canola oil, a 2020 Cochrane review found that eating less saturated fat led to a reduction in the risk of cardiovascular disease, including heart disease and strokes, and that health benefits arose from replacing saturated fats with polyunsaturated fats.[63]

The biggest problem with seed oils is how they are used. For example, oils are commonly found in unhealthy processed foods like biscuits and crisps, which tend to be high in salt, sugar and refined carbohydrates. They are also used for deep frying in takeaways and restaurants, which can be problematic, as heating oils to a high temperature and reusing them repeatedly can lead to toxic compounds building up. However, this isn't relevant to general use at home.

Another concern is hexane, which is a solvent used to extract oil from the seeds. However, the trace amounts of hexane that can be found in refined oils are below the safe limit guidelines, and hexane from all food sources reportedly makes up less than 2 per cent of the daily intake from all sources.[64] However, if you are worried, you can simply opt for cold-pressed and unrefined oil instead, as these are not extracted with solvents.

And importantly, all of this doesn't mean that seed oils necessarily need to be a part of a healthy diet. If you want to avoid them, you absolutely can. Despite this, the conversation around seed oils is filled with staggering amounts of misinformation and unsubstantiated assumptions. The fact that this is an argument that has been unquestioningly believed and reshared by so many people just goes to show how sadly true Brandolini's law really is.

'VEGANS NEED TO TAKE SUPPLEMENTS AND EAT FORTIFIED FOODS'

The criticism around vegans consuming supplements and fortified foods speaks to just how little we know about the food that we consume. Our entire food supply chain is filled with fortification and supplementation, even starting with our soils, which as well as being fertilised can be fortified with selenium and cobalt (see the B12 argument).

We fortify bread with iron, calcium and vitamin D. We fortify cereals with folic acid, B vitamins and calcium. We iodise salt and fortify fruit juices with calcium and vitamin D. We even fortify milk with vitamin D, iodine and vitamin A. Basically we fortify a wide range of foods and have done so for decades, thereby making it easier for people to get the nutrients they need.

We also fortify animal feed and give animals supplements, especially as the vast majority are factory-farmed. So even though we might talk about all the nutrients found in meat, the truth is many of those nutrients will have had their levels boosted or will only be there in the first place because of the fortification of the animal feed and supplements. This includes things like omega-3 for farmed fish, vitamin D and omega-3 for egg-laying hens, and nutrients such as selenium, zinc, vitamin E, vitamin B12, copper and iodine that are regularly added to animal feed or given to animals in the form of supplements. We even fortify the silage and hay that is used for grazing animals, and also give grass-fed, outdoor-raised animals supplements. So it's not only the intensively raised animals who have their diets enhanced with added nutrients.

Plus, even just looking at humans specifically, vegans are hardly the only ones keeping the supplement industry alive – supplement use is exceptionally common. Nobody bats an eyelid when someone says they take a fish-oil supplement, but a vegan taking an algae-oil supplement for the exact same reason is somehow seen as an indictment of veganism.

This argument is just fraught with double standards. Why is it wrong if vegans get nutrients from supplements but not if meat eaters do the same and eat animal products that have been given supplements and fortified foods in the farming process? In essence, vegans are just cutting out the bit where the nutrient gets filtered through an animal who is exploited and slaughtered and taking the supplement themselves instead.

But regardless of these double standards, what does it even matter? Let's say that you eat an animal product from an animal who didn't graze on fertilised or fortified soil, who never received a drench or a supplement and ate no fortified feed or silage. What exactly does that prove? That they deserve to die? Even in this hypothetical situation where it is only vegans who take supplements or eat fortified foods, how does that justify causing suffering, harm and death to animals? If we can take a supplement and remove our participation in the biggest cause of violence, cruelty and death for animals, the biggest driver in the emergence of bird and swine flu, the biggest user of antibiotics and the largest risk factor in the emergence of antibiotic resistance, and also in doing so eat a diet that is quantifiably and scientifically proven to be the most sustainable that we can eat, why on earth wouldn't we?

So even if we just ignore the myriad important reasons why this argument doesn't make sense, taking supplements doesn't invalidate veganism – it just means that we can take care of our health while also taking care of our planet and the animals. Ultimately, this argument is weaker than the bones the *Daily Mail* thinks vegans have.

So what supplements and fortified foods should a vegan consume? First, anyone who lives in a part of the world, like the UK, where there is less sunlight, particularly in autumn and winter, whether they are vegan or not, should be taking a vitamin D supplement. On top of that, vegans should be taking a vitamin B12 supplement, and I would also recommend an algae-oil supplement so that you get all the long-chain omega-3 fatty acids you need. Alternatively, it is possible to get a multivitamin specifically catered for vegans, which covers all your bases.

In terms of foods, opting for unsweetened and fortified plant milks is one of the easiest and simplest ways to make sure that you are getting a good serving of calcium and other vitamins, and actually many plant-based meat and cheese alternatives come fortified with nutrients like vitamin B12. You can also get fortified foods like nutritional yeast (which I promise tastes better than it sounds).

If in doubt, one of the easiest ways to find out if you are hitting all your targets is to use a food tracking app where you can log what you are eating and make sure you are getting all of your nutrient requirements.

'VEGAN DIETS ARE NOT HEALTHY FOR CHILDREN'

Won't somebody think of the children?

There's no surer way to turn people away from something than to claim it is bad for children. However, while this tactic can often be used cynically by people who have a vested interest in undermining veganism, it is nonetheless important to scrutinise whether a plant-based diet can be healthy for children too.

Stories about children dying who have vegan parents always make for viral news – for example, a story about an 18-month-old child with two vegan parents dying of malnourishment received international coverage in 2022. However, the parents only fed their child raw fruits and vegetables. A raw-food diet is not even optimal for an adult let alone a child, and the pseudo-beliefs that exist about raw-food diets are not the same as veganism.

In another high-profile case, a seven-month-old baby died because the health-store-owning parents diagnosed their child themselves as having a gluten and lactose intolerance and were giving them milk made from quinoa, buckwheat and rice, and visiting homeopathic doctors. (It's important to note that if breastfeeding isn't an option, there are dairy-free baby formulas widely available that are suitable for use from birth. There are also plant milks created specifically with children in mind.)

When actually looking into these stories, there is a recurring and obvious theme. It's not the fault of veganism if a parent is feeding their child an obviously nutritionally deficient diet that just so happens to be plant-based. These are both cases of clear and unacceptable negligence by parents who had unsubstantiated and anti-scientific beliefs. It wasn't veganism that killed these children – it was pseudoscience.

Using examples like these to claim that veganism is unhealthy for children would be like claiming that feeding meat to children is unhealthy because a child died due to their parents only feeding them raw animal organs. Using extreme examples of neglect, where parents were following pseudo-scientific beliefs, does not mean veganism is unhealthy for children, and it's disingenuous and fallacious to claim that it is.

So what about studies comparing vegan kids to non-vegan kids? One of the more salacious stories about these two groups came from a study looking at 187 children aged between five and ten years old in Poland. The vegan children in the study were 3cm shorter on average, leading to a whole host of news reports about veganism leading to stunted growth. However, one of the co-authors of the study said that the findings didn't mean the children were stunted in growth and could even grow to be taller than the meat eaters.[65] But why should we let an author of the study get in the way of a clickbait headline?

One finding of the study that was notable and important was that the vegan children had 5 per cent lower bone mineral density.[66] The data showed that the median calcium intake was lower than the recommended amount and there was inconsistent vitamin D or B12 supplementation across all of the vegan children, with nearly a third of the children on the meat-free diets not being given any B12 supplements or B12-fortified foods and only a third receiving vitamin D supplements. This information provides a rational explanation as to why the vegan children had lower bone mineral density on average.

On the flip side, the vegan children had the highest intakes of dietary fibre, folate, polyunsaturated fats, vitamin C, magnesium and more, and they also had better markers of cardiovascular health. The vegan children also had up to 81 per cent lower inflammatory markers, and the vegan children who were receiving B12 supplements had higher levels than those who ate meat. Shockingly, 30 per cent of the meat-eating children also had borderline or high LDL (bad) cholesterol.[67]

In other words, the study actually revealed both pros and cons for both diets, and the public perception of its findings was skewed by the way they were reported. If the headlines had read 'VEGAN CHILDREN HAVE BETTER CARDIOVASCULAR HEALTH AND HIGHER INTAKES OF HEALTHY FATS',

'VEGAN CHILDREN CAN HAVE BETTER B12 LEVELS THAN MEAT-EATING CHILDREN' or 'VEGAN CHILDREN FOUND TO HAVE HIGHER LEVELS OF CERTAIN ESSENTIAL NUTRIENTS' people would have felt more favourably about veganism.

That doesn't mean the negatives of the vegan diet shouldn't have been reported – it was an important takeaway that the vegans in this study had lower bone mineral density and lower intakes of certain nutrients. However, without a fair assessment of the findings overall, we were left with a biased summary that created the impression that a plant-based diet would inherently cause stunted growth and bone density loss. Plus a review of 437 studies concluded that a well-planned vegan diet allowed for children to grow normally and could even be beneficial.[68]

Proper planning is the key. While the research suggests that vegan parents need to properly plan so that their children get enough B12, vitamin D, calcium and other nutrients, it also suggests that non-vegan parents need to properly plan so that their children get enough dietary fibre, polyunsaturated fat, folate, magnesium and other nutrients.

More generally, other studies and dietary organisations, including the British Dietetic Association[69] and the Academy of Nutrition and Dietetics[70] among many others,[71] [72] have concluded that there are no nutritional risks to children following a vegan diet when due attention is given to B12 and vitamin supplementation and calcium intake, with a plant-based diet being labelled as 'appropriate for all stages of the life cycle.'[73]

Being a parent means looking after your children's needs, and ensuring they are safe and getting all of the nutrients they require from the food they eat. This applies to both vegan and non-vegan parents. Just as a non-vegan diet doesn't necessarily mean a child is getting all the nutrients they need, a vegan diet is the same.

So it's not really a question of whether a vegan diet is healthy for children, because it can be. It's about a parent making sure their child is getting everything that they require, which any parent should be doing regardless of dietary preference. Moving forward, if we really cared about the safety of our children, we would move away from scaremongering and the dissemination of misinformation, and look to support parents and caregivers raising their kids as vegans, rather than accusing them of neglect or abuse, as is sadly sometimes the case.

'PROCESSED MEAT ALTERNATIVES ARE WORSE THAN MEAT'

There has been no shortage of headlines over the last couple of years declaring that plant-based alternatives are actually really bad for you. One even stated: 'WHY VEGAN MEAT SUBSTITUTES CAN BE WORSE FOR YOUR DIET THAN JUNK FOOD.'[74]

I'll delve into these more conventional claims in just a moment, but my favourite accusation about what's in plant-based alternatives was put forward by the American rapper T.I. Harris when he was talking on his podcast with renowned 'intellectual' Alex Jones. Harris said:

A lot of people [are] wondering, *Why does this Impossible Meat taste so much like meat?* And it is a conspiracy theory ... that there may be human meat in the Impossible Burger. Real meat and human meat, and they say that a lot of the derelicts, homeless people and other missing cases that have gone cold could be the basis of the Impossible Burger.

Unfortunately, beliefs like these are not just isolated to this podcast episode. The US-based Newsmax channel aired a programme in which one of the producers of the show stated that they wouldn't eat an Impossible Burger because they were worried that it was 'made of chemicals, implanted with chips ... I don't want to eat Bill Gates's plate of fake meat.' She was quite right, of course, to be worried about chips being implanted in a burger, as we all know that chips belong on the side, preferably with a tasty plant-based condiment. Needless to say, there are no microchips or human flesh in the Impossible Burger, or any other plant-based alternative for that matter.

But what about the more reasonable questions around the general healthfulness of plant-based alternatives to meat? The first thing to recognise is that plant-based alternatives are not meant to represent the zenith of health and well-being. And they're not meant to replicate the healthfulness of fruits and vegetables – they're meant to replicate the taste, texture and satisfaction of meat products.

Criticising vegan bacon for not being a health food is a straw man argument, as vegan bacon isn't meant to be a health food. Plus, you don't have to eat these alternatives as a vegan – they're not a requirement. The vegan police don't turn up and take away your vegan card if you've not had a plant-based burger in the past 72 hours. (Although if they did, microchips would definitely make that process easier, Bill, if you're reading.)

The phrase 'plant-based alternative' is also very generic. Tofu can be used as an alternative but is very healthy. However, there are some processed plant-based alternatives that can contain unhealthy levels of things such as salt and saturated fat. Obviously this means that those foods should be eaten in moderation from a health perspective or simply avoided if you so choose. The fact that certain alternatives are not healthy foods in their own right does not, however, mean we should eat animal products instead. 'Oh, this vegan bacon is high in salt. I guess it's just the class 1 carcinogen bacon for me then.'

That being said, I do strongly believe that plant-based companies should strive to make their products healthier; however, they find themselves stuck between a rock and a hard place. On the one hand they are criticised if they don't taste like the processed meat products they're replicating, something that a lentil-and-kale burger will of course never be able to do. But then they're criticised when they're processed so that they can actually convincingly replicate meat products. So be healthy and be criticised for not satisfying the taste meat eaters want, or satisfy what meat eaters want and not be the healthiest foods.

Hopefully, plant-based alternatives will become healthier over time as the technology improves and the sodium and saturated fat content is reduced, but how do they compare health-wise to their meat counterparts now? A study looking at the consumption of Beyond Meat versus organic grass-fed beef found that consuming the plant-based alternative lowered TMAO levels (a risk factor for cardiovascular disease), lowered LDL cholesterol and meant less saturated fat was consumed while also increasing fibre consumption.[75] The study did receive funding from Beyond Meat, but the data analysis was conducted by a third party that wasn't involved in the design of the study or the collection of the data and was blinded to all study participants.

Another study found that plant-based alternatives increased microbiome health compared to their meat counterparts.[76] And an analysis of 207 plant-based alternative products and 226 animal products in the UK found that the plant-based products 'have a better nutrient profile compared to meat equivalents', as well as lower energy density, less saturated fat and significantly higher fibre. They did, however, also have higher levels of salt. The authors stated that the study supported the claim that plant-based meat 'is a healthier alternative to animal products from a chronic disease prevention perspective'.[77]

So where does that leave us? Ultimately, processed plant-based alternatives are not the healthiest foods that we can consume, nor should they be viewed as such. They do, however, appear to be preferable to their meat counterparts, including even grass-fed and organic red meat. And, as I said, hopefully they will become even healthier in the future.

But it is also important to emphasise that veganism is not about health, although there can of course be real health benefits to a plant-based diet. If the merits of these alternatives are judged from the perspective of eating vegan sausages so that we stop paying for gas chambers, that is an irrefutable point, even if said vegan sausage is not objectively healthy.

And that's exactly why I do eat vegan sausages from time to time – because I like how they taste and by eating vegan sausages I'm not paying for pigs to be forced into gas chambers. I know it's not as healthy as having a chickpea-tofu-and-sweet-potato stew, but if I wanted the health benefits of that meal, well, that's what I would eat.

'THERE ISN'T ENOUGH EVIDENCE TO SUPPORT THE CLAIM THAT VEGANISM IS HEALTHY IN THE LONG-TERM'

There are certain places around the world that are notable due to the high average life expectancy of the people who live there. These are called blue zones, and there are five of them: Okinawa, Japan; Sardinia, Italy; Nicoya, Costa Rica; Ikaria, Greece and Loma Linda, California.

Even though these areas are spread around the globe, when the lifestyles of the people who lived in these areas were studied, they showed some very interesting commonalities. They all have a focus on being social and existing within communities, they practise stress management, and they find purpose in what they do. They are also active communities – this means things like gardening, growing food and community-based manual labour.

However, while all of these seem fairly self-explanatory and are certainly not surprising, the most important aspect of what they all share in common is their diet. Blue zone regions eat a 95–100 per cent plant-based diet. And within this diet they prioritise foods like legumes, nuts, seeds, greens, fruits, wholegrains and potatoes.

Those who do eat animal products do so sparingly, and in the words of one researcher who has studied blue zones, 'We don't know if they lived longer despite eating meat.'[78] In fact, a study of people in Loma Linda has shown that within that community, the longest living were those who followed a plant-based diet and those who ate a plant-based diet with a small amount of fish included as well.

So actually the evidence suggests that a plant-based diet can absolutely ensure a long and healthy life and might even be ideal for doing so, especially if combined with other important lifestyle factors such as movement, socialising and living with purpose.

And it's not just the age that people in blue zones live to that is notable – it is also their mobility. People who live in these regions have higher rates of mobility and are more active at an older age. After all, health is not simply about how long you live but also about the quality of life itself.

'VEGANISM CAUSES EATING DISORDERS'

Eating disorders have long been used to criticise veganism, with some people arguing that because veganism is by its nature a selective way of eating, it can lead to disordered eating, centred around restriction and a heightened awareness of what is in food. However, the issue of restricting food is not as simple as that. After all, no longer consuming sugary soft drinks is a form of restriction, but it is by no means a bad thing to do or suggestive of an eating disorder. So not all restriction is necessarily unhealthy or indicative of a wider problem.

People who are looking to restrict calories might naturally look to plants as a better way of doing so, as plants generally have fewer calories than animal products. And I have heard anecdotes about people who started eating plant-based doing so because they wanted to mask their disordered eating or because doing so fitted into the mentality their disordered eating had given them about food. However, at the same time, I have also heard anecdotes about people who found that veganism actually helped them to recover from an eating disorder and gave them a healthier relationship with food. So when it comes to people who have a history of disordered eating, there is no hard-and-fast rule.

Obviously people shouldn't eat plant-based to mask an eating disorder. However, if someone who has struggled with disordered eating genuinely wants to become vegan, and as a consequence make dietary changes, the key is to find the best way to manage that decision while ensuring the person is looking after their mental and physical health. For such a person, making changes over time rather than simply removing all animal products from day one could well be a safer and more sensible option. Disordered eating comes in different forms, and the experiences of those who have battled with it will also be different and varied.

But does going vegan cause disordered eating?

In the biggest study of its kind looking specifically at this issue, vegans actually scored significantly lower than omnivores on the Eating Disorder Examination, which measured pathological eating behaviour.[79] This means that the vegans in the study exhibited fewer eating disorder tendencies and were shown to have healthier attitudes towards food. They also scored lower on things like food restraint, food concern and on the binge eating scale as well.

Plus, looking at the evidence regarding this issue led to the largest body of diet and nutrition professionals in the world stating that 'eating disorders have a complex etiology and prior use of a vegetarian or vegan diet does not appear to increase the risk of an eating disorder'.[80]

'VEGANS ARE MORE DEPRESSED'

Towards the end of 2022, a study on people in Brazil was published that claimed that those who didn't eat meat had twice as many depressive episodes as omnivores.[81] Considering the number of criticisms we have to respond to and jokes about veganism being a big 'miss-steak', can you really blame us? Funnily enough, anti-vegan jokes haven't evolved at all in the time I've been vegan. It's almost as if meat eaters are stuck in the past or something.

In all seriousness, though, is there any evidence that vegans are more prone to depression?

The study just referenced is not a strong one. Out of the 14,000 people surveyed, only 82 didn't eat meat, or fewer than 0.6 per cent. Making assertions based on such huge discrepancies in sample sizes is going to produce tenuous results at best. Furthermore, the researchers also ruled out factors such as nutritional deficiencies and lifestyle factors, rendering diet almost irrelevant to their findings.

However, this isn't the first study that found an association between mental health and abstaining from meat. A 2020 review that was 'funded in part via an unrestricted research grant from the Beef Checkoff, through the National Cattle-men's Association' claimed that avoiding meat was associated with higher rates of depression, anxiety and/or self-harm.[82]

One of the biggest issues with this review was that the majority of the studies they included were cross-sectional, meaning that unlike longitudinal studies that follow partici-pants over a period of time with repeated monitoring of risk factors and health outcomes, the studies instead looked at a single point in time to evaluate the variables. This is espe-cially relevant when discussing mental health, as evaluating such conditions requires a consideration of a person's history and working with them over a period of time.

The highest quality evidence comes from randomised con-trolled trials, and the only randomised controlled trial featured within the review actually stated that 'vegetarians reported significantly better mood than omnivores and fish eaters after the trial'. Of the studies that actually looked at and included vegans, four of them showed no difference between vegans and non-vegans, one showed that vegans were happier and two showed a negative association but were not statistically significant or the sample size of vegans was so small they didn't look at them as their own group.

Other studies included in the review that focused on vegetarians versus meat eaters did find more of a negative association between meat abstinence and mental health. However, the reason why such an association was found was not explored by the authors, nor were important factors accounted for.

One such factor would have been gender differences. While younger females are more likely to not eat meat, they are also more likely to suffer from mental health conditions such as depression – much in the same way that adult males have the highest rates of suicide but are also the most likely to eat meat. Does that mean that eating meat is linked to male suicide, or does it mean that there are social, cultural and gender factors that need to be acknowledged? For example, eating meat is seen as being manly, as is being stoic and unemotional or not asking for help.

There's also the issue of reverse causation, whereby people who do suffer from mental illness might be more likely to make dietary changes in a bid to try and influence their mental health positively. For example, many people who struggle with their mental health take up some form of exercise to help them or try a practice like meditation. A cross-sectional study could then find higher rates of mental health conditions within a group of people who have taken up meditation, for example, versus a group who have never tried it. However, such a finding would not show that meditation causes depression or anxiety.

There is also evidence that emotional empathy is correlated with depression, particularly among adolescents,[83] while at the same time those who choose not to consume meat have been associated with greater openness and empathy.[84] This suggests that even if an association was made between not eating meat and depression, the two might be linked because of increased emotional empathy, rather than because the lack of meat has caused depression.

Even from a more common-sense perspective, caring about issues of social justice and harm means exposing yourself to upsetting and challenging imagery and realities. This is why the issue of 'climate anxiety' within the environmental movement is becoming increasingly spoken about. Does holding a placard at an environmental rally cause mental health problems, or does becoming emotionally invested in the currently worsening future of our planet and the suffering that is being caused as a consequence of it negatively impact an individual's mental health?

The same is true of veganism. Does not eating meat cause mental health issues? Or is your mental health adversely affected by being emotionally invested in the suffering of tens of billions of animals, watching footage of them being abused, while at the same time realising that the vast majority of people in the world are paying for these things to happen to them, including your friends and family?

Importantly, there have also been other studies that have shown that removing meat from your diet is beneficial to mental health, such as one in which employees of a major insurance company were split into a vegan group and a control group who just ate as normal. In the study, the vegan group demonstrated improvements in general and mental health compared to the control group.[85]

Undeniably, avoiding nutritional deficiencies is important when it comes to mental health; however, such deficiencies can be avoided on a plant-based diet. Ultimately, attempting to claim that not eating meat will make you more depressed is not based on anything mechanistic and is a reductive conclusion made by looking at certain studies while ignoring others, and failing to account for important factors and variables. Mental health is an issue of huge importance, and for the meat industry to fund bad science to create poorly substantiated associations and promote eating meat is unconscionable.

THE PRACTICALIST

'VEGANISM IS INCONVENIENT'

Due to the ubiquity of animal products and the fact that we need to change the types of food that we eat, veganism can be an inconvenience in some situations. However, do we rank the morality, or indeed the importance, of something based on its convenience? After all, lots of things that are important and beneficial are inconvenient.

The reality is, most of the inconvenience of veganism goes away once we've been vegan for a little while. The biggest adjustments that we have to make at the beginning are to the recipes we cook, finding out what plant-based foods we like and checking labels for certain products. However, once we've got into the flow of being vegan it becomes our new normal.

Sure, we might have to check that the restaurant we're going to with our friends has a vegan option, and when travelling to places outside of cities where veganism is less prevalent, we might need to plan ahead a little, but how do these things stack up compared to the harm that buying animal products causes?

It's also pleasing to say that veganism is becoming more convenient all the time. When I first went vegan in 2015, very few restaurants had good vegan options, and my local supermarkets had no plant-based alternatives. This is no longer the case, and things are improving all the time. Also, by embracing the inconvenience that this change in diet initially presents, not only are we having a positive impact on the lives of others through our food choices, but by normalising vegan food and supporting vegan products, we are making veganism more convenient for people in the future as well.

'I DON'T HAVE TIME TO BE VEGAN'

I can absolutely see a situation where this could be true – for example, a single mum who's working overtime to be able to stay afloat might be reliant on grab-and-go food, except she lives in an area where there are only a limited number of options to begin with, let alone vegan ones.

However, while I acknowledge that there are practical barriers to some people being able to have food autonomy, this is not the situation that the majority of us find ourselves in. If we already make our own breakfasts and cook our own meals, we have time to be vegan because we're cooking and making food anyway. Cooking plants doesn't take any longer; in fact, it can actually take less time. And if we buy our food from supermarkets or order it online, we can continue to do so but just buy plant-foods instead.

Of course, some time might need to be invested when you first become vegan as you research vegan recipes and learn how to get all of the nutrients you need, but once you have done that, it's done. The benefit of taking that bit of time at the beginning to properly prepare yourself means that you can then live in a way that is more sustainable and more ethical.

Again, just like with the small amount of inconvenience that can be caused when we become vegan, a bit of extra time at the start when we go vegan is nothing compared to the harm that is caused if we don't.

'VEGANISM IS TOO EXPENSIVE'

This is an exceptionally common criticism, but is it actually true? Well, yes and no. I think of it this way: if someone eats meat, they could go to a high-end steakhouse and spend inordinate amounts of money on a piece of cow flesh, or they could go to a fast-food chain and spend very little.

The same is true for veganism. You can absolutely spend a lot of money as a vegan and buy lots of artisanal and organic cashew-based cheeses, but you don't have to do that. In other words, you can make veganism work on any budget.

Now, it is true that plant-based alternatives like burgers and cheeses can be more expensive than their meat and dairy equivalents, but that isn't necessarily always the case, and it should also change over time as demand for alternatives increases. However, the fact that this argument is normally made with those kinds of products in mind overlooks what the core of any healthy plant-based diet should be: whole-plant foods. These are foods like fruits, vegetables, wholegrains, beans, lentils, nuts, seeds, and so on.

A study from the University of Oxford published in 2021 looked at the costs of different diets. Pescatarian diets were shown to be the most expensive and vegan diets that consisted of whole-plant foods were shown to be the most affordable. In fact, in high-income nations, wholefood plant-based diets were shown to reduce food costs by up to one third compared to the standard omnivorous diet.[1] So it's not the case that veganism is too expensive, it's just that some products can be more expensive, but overall it can actually be more affordable.

Miguel Barclay, who is known for his #OnePoundMeals, supports this claim: 'I definitely agree that cutting down your meat, or cutting it out completely, will save you money …

without doubt vegan and vegetarian meals consistently come in at a much lower price than recipes with meat.'[2]

There are several factors that influence the prices of the foods that we eat. One of the most important of these is the profit margin added to food prices by the supermarkets. In the Netherlands, for example, meat is commonly marked up with an 8 per cent margin, while plant-based alternatives can have margins of around 35–50 per cent.[3] If supermarkets were to apply the same margins to plant-based products as they do to animal products, there would be much closer price parity. And, in fact, this has started to happen: the UK supermarkets Tesco and Co-op have both reduced the margins on their own-brand plant-based products to match the price of the animal-based foods they're replicating.

There's also all of the externalised costs associated with food. These include healthcare costs and environmental clean-up costs. A 2020 study concluded that conventional meat prices in countries in the EU would need to increase by almost 250 per cent if externalities such as greenhouse gasses and land-use change were taken into account. For conventional dairy, it would be a 91 per cent increase in price; for comparison, plant-based products would only require a 25 per cent increase.[4] This also doesn't include the associated healthcare costs, which on their own would make processed meat cost 110 per cent more.[5]

Another report, looking at the hidden costs of our global food system, estimated that the price of these externalities is somewhere in the region of $12 trillion a year (more than the value of the food produced), with that figure estimated to rise to $16 trillion by 2050. The same study also showed that switching our food system to one that adopts the human and planetary health diet, which is mostly plant-based, would only cost $30 billion annually but would result in economic benefits of $1.28 trillion annually as well.[6]

The price of animal products also reflects what animal

farming looks like. After all, there's a reason why a whole chicken costs so little to buy. The intensification of animal farming coupled with practices such as selective breeding and antibiotic use mean that we raise tens of billions of animals in environments where they are crammed in beak-to-beak and snout-to-snout. These animals reach slaughter weight at a fraction of their lifespans – six weeks in the case of chickens, and their lives are filled with suffering, as they are viewed as commodities and farmed in such a way as to maximise the productivity and profitability of their flesh and secretions. In other words, animal products are also cheap because of the huge ethical costs associated with their consumption.

Then there are agricultural subsidies, public funds raised through taxation that are given to the farming sector. According to the UN Food and Agriculture Organisation, around $540 billion is given to farmers every year,[7] with the biggest subsidies going to the production of animal products and animal-feed crops. Subsidies can be a great way of ensuring that consumers have access to accessible and affordable food, but in their current form agricultural subsidies are propping up unethical and unsustainable systems of farming. If we were to redistribute public funds so that they weren't incentivising animal farming but instead plant farming, rewilding and food innovation, such as precision fermentation and cell-cultured meat, this would mean we could actually produce more food and make it more affordable and accessible, further driving down the cost of being vegan.

'NOT EVERYONE IN THE WORLD CAN GO VEGAN'

Due to factors like poverty or food scarcity, this is of course absolutely true. And, as discussed in 'The Well-Intentioned Leftist', for some people in the world, a plant-based diet won't be possible. Even in Western countries, people reliant on food banks won't necessarily be able to eat only plants. So there are some people who do genuinely lack the ability to be vegan.

If you can't be vegan, you can't be vegan. However, the only relevant question for each individual is: can *I* be vegan? It doesn't matter what other people can or can't do – what matters is what each of us can do.

When we use this argument, we are appropriating other people's situations to try and find a justification for ourselves. However, the fact that not everyone in the world can go vegan actually provides those of us who can with even more of an incentive to make that change. Continuing to support the same system means we'll continue to be given the same system. So if our argument against veganism is one around accessibility, then the best way to make veganism more accessible and our food system more equitable is to be vegan and actively support it so that we signal to companies and legislators that that's what we want.

A large part of the accessibility argument is that people believe veganism means eating plant-based alternatives to meat, which can be harder to find and, as previously mentioned, more expensive. But eating plant-based alternatives to meat is not what veganism is. Of course they are vegan, but you can be vegan and not eat meat-free burgers and sausages.

People sometimes say to me, 'I don't like vegan food,' which means they don't like fruits, vegetables, whole-grains, potatoes, nuts, lentils, beans, and so on. Vegan or not, there's a pretty high chance you're eating food that is suitable for vegans most days, if not every day. And if we buy our groceries in a supermarket, then veganism is very accessible, regardless of whether or not they have a good range of plant-based alternatives. When we reshape what we perceive eating vegan to be, all of a sudden there are lots of plant-foods for us to enjoy.

'MY FAMILY WON'T LET ME BE VEGAN'

It is certainly the case that living with family, and the lack of financial and general independence that often comes with that, means that there will most likely be more barriers to making the change to veganism than there is for someone who has more independence and autonomy.

For anyone who does want to be vegan but can't because they still live at home and their family members don't support that decision, the solution is to not let the difference of opinion cause any rifts or destabilise any familial relationships. Instead, perhaps see if your family would be happy to try some vegan products or to adapt some of their recipes to be vegan. You could even offer to cook some vegan meals or, if you have your own money, offer to buy in some plant-based foods for your family to try and see if they like them too. Then, if you haven't been able to go vegan before you leave home, when you actually do, you can make that choice.

Now, for some people living at home, it might not be quite as simple as offering to cook some meals yourself, and for some people the resistance from their family might be so much that they won't even entertain any of the suggestions I just mentioned. In those cases, it's just a matter of doing what you can, eating plant-based when it's possible but also not burdening yourself with an unrealistic expectation to do something you are not able to.

That being said, I do think that in cases where our family's reaction might be upsetting, it's a good idea to try to understand why they are reacting that way. For some parents it might be a case that they are worried about you – perhaps they are concerned you won't get all the nutrients you need to be healthy. While it could be hurtful, this worry comes from a good place and is about a parent wanting to make sure their child is healthy and well. If this is the case, take the time to find out where you can get all the nutrients your parents are concerned about, so at least if they bring it up, you can attempt to allay their concerns.

In other situations, a parent might not seem to be concerned and could even appear to be acting spitefully or in a mean way. It could well be that such a parent feels like they are having their parenting and indeed their character attacked by the choice their child is making. After all, as well as making sure their children are safe and healthy, one of the other most important jobs of a parent is to make sure their child acts in an ethically responsible way. If their child then turns around to them and says they're going vegan because what we do to animals is immoral, they're essentially saying that the parent raised them to do immoral things.

I of course recognise that it isn't quite as simple as this – there is a lot of nuance around eating animal products due to cultural norms, the way animal products are marketed, and so on. So while a child won't necessarily be attacking their parent – after all, their parent is merely raising them the way they were raised and the way the majority of people are raised – it is understandable why a parent might feel personally offended. Perhaps they then act in a mean way because they want their child to stop being vegan so that they can remove the feelings of hurt their child has caused them. This situation is undeniably very hard, so I would suggest just trying your best to not take it personally and remember the reasons you have become vegan. Hopefully, over time, your parents will get used to your choice and it will become easier as a result.

Attempting to have an understanding of why people around us act the way they do is an important aspect of working out how to solve any potential issues or frictions that are arising. I also think it's important to get to the bottom of what our family members' concerns and objections really are: perhaps they're asking about nutrients, perhaps they're asking about recipes and cost, or perhaps they're arguing that if it wasn't for our ancestors eating meat, we wouldn't be around today. Whatever it is that they are saying, I think having a response prepared is a good way of showing your commitment to and seriousness about veganism, and it will also hopefully bring them around to the idea over time as their objections are tackled one by one.

'NONE OF MY FRIENDS ARE VEGAN, SO IT WOULD BE AWKWARD FOR ME TO BE VEGAN AROUND THEM'

When it comes to friends, the easiest way to tackle this problem is just to communicate with them that you have gone vegan so you don't surprise them or make it awkward by telling them when you are about to go out for food, or even worse, when they've cooked for you.

Then, if you are going out for a meal, you can make sure in advance that the restaurant has a vegan option. It can make it more awkward if your friends want to go somewhere that doesn't have a vegan option, but it is so much easier to find vegan food now than it used to be. Sometimes restaurants actually have vegan options that aren't labelled as such, and sometimes you can make a meal vegan by just removing one or two ingredients from it. So while it is possible that your veganism could make certain social situations a little more awkward, it becomes easier and easier over time.

Sometimes friends can mock or tease you for being vegan, but the important thing is to remain confident and assured in the decision you've made. As with unsupportive family members, just take some time to educate yourself about what it is that they are saying so that you have a good response.

'I DON'T KNOW HOW TO COOK'

If you don't know how to cook, perhaps becoming a vegan offers you the perfect chance to learn. Cooking vegan meals is no more complicated or time-consuming than cooking non-vegan meals. In fact, it can often be easier, as you don't have to worry about food poisoning from raw meat.

If it is that you simply don't want to cook, then there are more and more vegan ready meals becoming available all the time. You can also order vegan food via takeaway apps, and there are easy plant-based foods that you can just cook in the oven. That said, from a health and cost perspective, whether you are vegan or not, learning how to cook just a handful of recipes is a great thing to do. There are so many plant-based cookbooks and recipe accounts on social media. If you want a vegan version of a meal that you've always liked, then you can just google it and put the word 'vegan' at the start. You can find vegan recipes for more or less every meal that you can think of these days.

Veganism also offers us the chance to explore new flavours and ingredients. While we might view it as an inconvenience or a hassle to learn to cook new foods, it can actually be a joy. It can be exciting to find out how to 'veganise' our favourite dishes, and it is often just as simple as replacing one or two ingredients from the recipe we would normally cook anyway.

I think of veganism as not being about reinventing the wheel – it's more like we are just adding some new shiny alloys to the wheels we already have. It's the same thing, just slightly fresher and newer.

AFTERWORD

We patronize them for their incompleteness, for their tragic fate for having taken form so far below ourselves. And therein do we err. For the animal shall not be measured by man. In a world older and more complete than ours, they move finished and complete, gifted with the extension of the senses we have lost or never attained, living by voices we shall never hear. They are not brethren, they are not underlings: they are other nations, caught with ourselves in the net of life and time, fellow prisoners of the splendour and travail of the earth.

Henry Beston, *The Outermost House: A Year of Life on the Great Beach of Cape Cod* (1928)

What is it like to be an animal? We might automatically think of a reduced experience, one of comparative simplicity and intellectual naivety. However, it's ultimately impossible to know.

While humans can comprehend complex issues that other animals simply have no capacity to understand, we fall short of comprehending the experiences that they have. After all, what is it like to focus on an object two miles away, as it is believed that eagles can? What is it like to be able to detect smells 12 miles away or to hear at frequencies 20 times lower than humans, as it is believed elephants can? What is it like to have 10 times as many tastebuds as we have, which is the case for catfish? What does it feel like to run at 70 miles per hour, like a cheetah, or dive to depths of more than 2,000m in the ocean like a sperm whale?

Einstein once said, 'Everybody is a genius. But if you judge a fish by its ability to climb a tree, it will live its whole life believing that it is stupid.' And if you judge a human's ability by how far we can see, what we can hear and how well we can smell, we too end up falling short.

Even in the case of the animals we eat, animals we regularly demean and devalue intellectually, they too have experiences that we can't comprehend. When salmon migrate, they are able to find their way back to the rivers they were born in. It is believed that they navigate using the Earth's magnetic field like a compass and their sense of smell. It is believed that cows can detect smells up to 6 miles away and are able to hear frequencies that we are unable to. What does a slaughterhouse smell like to a mother cow crammed into a truck that is driving her to her death? The smell must get stronger and stronger as she is taken closer and closer to the bloody knife waiting to be pulled across her throat.

One of the most notable things in a slaughterhouse is the stench. A concoction of festering organs and blood, it smells like death. What is such a stench like to a cow, who has everything to fear from such an environment?

And if animals can hear, see, smell and taste more than we can, what is to say that there are animals who can't feel more than we can? If an animal can see greater distances, can an animal not feel greater pain?

As Henry Beston stated, we view animals as having taken form so far below ourselves. However, we do this by measuring them against what we can experience and achieve. There is no denying that humans have cognitive capabilities that other animals do not. There is no denying that we understand things that other animals cannot comprehend. And yet there is so much that we cannot comprehend either.

Throughout this book I have responded to as many arguments against veganism as I could, although I'm sure a particularly staunch meat eater will still manage to get creative and think of something that I have omitted. 'What if aliens came down to Earth and forced you into an alternate reality where plants were high-functioning and sentient members of society and chickens were insentient automatons filled with a delicious and nutritious candy-floss-type substance. Would you eat animals over plants then?' To be fair, that's not too far off something Neil deGrasse Tyson might say.

However, while I have provided a rebuttal to each and every one of these arguments, the strongest argument in favour of veganism is not to be found in the words in this book but from the animals themselves. They too are experiencing the world around them. They too are prisoners of the splendour and travail of the earth.

The world is undeniably a merciless and cruel place at times, and if there's one thing nature's not short of, it's suffering. However, one of the things that makes us unique as humans is our ability to display autonomy and self-determination in ways that other animals, still confined within the prism of the natural world and without the ability to scrutinise or change their actions, are unable to. We can make choices, critically reflect upon and challenge our behaviours, improve ourselves, and seek to make the world around us a better place. Although it is unlikely that the world will ever be devoid of harm and suffering, should we not strive to get as close to that aspiration as possible?

A question I often pose to people is: let's say that you were presented with a blank canvas and that you could design what our food system looked like and how we interacted with other animals – would you design what we have today?

Would you design a world with rampant deforestation and environmental collapse a mere moment away? Would you design a world with slaughterhouses? Would you design a world with so much needless animal suffering and death?

If you wouldn't, what would you come up with instead? Perhaps a world where animals didn't need to be slaughtered for food. Perhaps a world where we didn't need to cause so much devastation to feed ourselves? Perhaps a world where suffering, fear and harm were minimised and eradicated wherever possible?

If this is the world we would design, then why is this not the world that we strive to create now? This is not a world that can only be achieved through a blank canvas; it is a world we can fundamentally work towards creating by making different choices today, tomorrow and the next day.

I hope that I have managed to respond to and debunk any arguments you might have heard used against veganism, and I hope that you now share the belief that I have, if you didn't already, which is that something must change. And that change might be as simple as aligning our actions with our values. After all, the arguments for veganism are not ones that people don't necessarily already agree with. We all want a food system that is more equitable, affordable, accessible and healthier. We all want a food system that reduces the emergence of chronic diseases, infectious diseases and antibiotic resistance. We all want a food system that is more sustainable, more harmonious with the natural world, and better for the planet that we live on. And we all want a food system that is more ethical, more moral, and creates less suffering and harm.

These are things we all want, and it just so happens that we can bring them about by changing how we eat. Veganism is not the solution to all of the world's problems, but there is no solution to all the world's problems without it.

I hope that by covering all the main arguments against veganism that have been levelled against me, you now feel confident, informed and equipped to have your own conversations. I also hope that this book can serve as a tool for you and that you can refer back to it and use it to help you deal with any arguments against veganism that you might face in the future.

Martin Luther King Jr. once said: 'Our very survival depends on our ability to stay awake, to adjust to new ideas, to remain vigilant and to face the challenge of change.'

So will we face the challenge of change? For the sake of ourselves, our planet and of course the animals who are caught in the net of life and time with us, I hope the answer is yes.

NOTES

The Mistaken Philosopher

1. Reeve, C., 'Farm business incomes increasingly reliant on direct payments', AHDB (7 January 2021): https://ahdb.org.uk/news/farm-business-incomes-increasingly-reliant-on-direct-payments
2. Benton, T. G., Bieg, C., Harwatt, H., Pudasaini, R. and Wellesley, L., *Food system impacts on biodiversity loss*, Chatham House (2021): https://www.chathamhouse.org/2021/02/food-system-impacts-biodiversity-loss/01-introduction
3. Ritchie, H., Rosado, P. and Roser, M., 'Environmental impacts of food production', Our World in Data (2022): https://ourworldindata.org/environmental-impacts-of-food?insight=food-plays-a-large-role-in-many-environmental-impacts

The Wishful Thinker

1. 'Numbers of farmed fish slaughtered each year', fishcount.org.uk: http://fishcount.org.uk/fish-count-estimates-2/numbers-of-farmed-fish-slaughtered-each-year
2. 'Estimated numbers of individuals in annual global capture tonnage (FAO) of fish species (2007–2016)', fishcount.org.uk: http://fishcount.org.uk/studydatascreens/2016/numbers-of-wild-fish-A0-2016.php
3. Mackay, M., Hardesty, B.D. and Wilcox, C., 'The intersection between illegal fishing, crimes at sea, and social well-being', *Frontiers in Marine Science* 7 (2020): https://www.frontiersin.org/articles/10.3389/fmars.2020.589000/full
4. Amos, I., 'Salmon fatalities at fish farms "double" in past year, figures suggest', *Scotsman* (16 January 2023): https://www.scotsman.com/news/environment/salmon-fatalities-at-fish-farms-double-in-past-year-figures-suggest-3987766
5. Sneddon, L.U., 'Pain perception in fish: evidence and implications for the use of fish', *Journal of Consciousness Studies* 18(9) (2011): https://www.wellbeingintlstudiesrepository.org/cgi/viewcontent.cgi?article=1038&context=acwp_vsm
6. Sneddon, L.U., 'Evolution of nociception and pain: evidence from fish models', *Philosophical Transactions of the Royal Society* 374(1785) (2019): http://doi.org/10.1098/rstb.2019.0290
7. Ma. D.F., Qin, L.Q., Wang, P.Y. and Katoh, R., 'Spy isoflavone intake increases bone mineral density in the spine of menopausal women: meta-analysis of randomized controlled trials', *Clinical Nutrition* 27(1) (2008): https://pubmed.ncbi.nlm.nih.gov/18063230/

303

8. Okekunle, A.P., Gao, J., Wu, X., Feng, R and Sun, C., 'Higher dietary soy intake appears inversely related to breast cancer risk independent of estrogen receptor breast cancer phenotypes', *Heliyon* 6(7) (2020): https://www.sciencedirect.com/science/article/pii/S2405844020310720

9. Reed, K.E., Camargo, J., Hamilton-Reeves, J., Kurzer, M. and Messina, M., 'Neither soy nor isoflavone intake affects male reproductive hormones: an expanded and updated meta-analysis of clinical studies', *Reproductive Toxicology* 100 (2021): https://doi.org/10.1016/j.reprotox.2020.12.019

10. Martinez, J. and Lewi, J.E., 'An unusual case of gynecomastia associated with soy product consumption', *Endocrine Practice* 14(4) (2008): https:///doi.org/10.4158/EP.14.4.415

11. Messina, M., 'Soybean isoflavone exposure does not have feminizing effects on men: a critical examination of the clinical evidence', *Fertility and Sterility* 93(7) (2010): https://doi.org/10.1016/j.fertnstert.2010.03.002

12. Giampietro, P.G., Bruno, G., Furcolo, G., Casati, A., Brunetti, E., Spadoni, G.L. and Galli, E., 'Soy protein formulas in children: no hormonal effects in long-term feeding', *Journal of Pediatric Endocrinology and Metabolism* 17(2) (2004): https://doi.org/10.1515/JPEM.2004.17.2.191

13. Chavarro, J.E., Toth, T.L., Sadio, S.M. and Hauser, R., 'Soy food and isoflavone intake in relation to semen quality parameters among men from an infertility clinic', *Human Reproduction* 23(11) (2008): https://doi.org/10.1093/humrep/den243

14. Mínguez-Alarcón, L., Afeiche, M.C., Chiu, Y.-H., Vanegas, J.C., Williams, P.L., Tanrikut, C., Toth, T.L., Hauser, R. and Chavarro, J.E., 'Male soy food intake was not associated with in vitro fertilization outcomes among couples attending a fertility center', *Andrology* 3(4) (2015): https://doi.org/10.1111/andr.12046

15. Beaton, L.K., McVeigh, B.L., Dillingham, B.L., Lampe, J.W. and Duncan, A.M., 'Soy protein isolates of varying isoflavone content do not adversely affect semen quality in healthy young men', *Fertility and Sterility* 94(5) (2010): https://doi.org/10.1016/j.fertnstert.2009.08.055

16. Applegate, C.C., Rowles, J.L., Ranard, K.M., Jeon, S. and Erdman, Jr., J.W., 'Soy consumption and the risk of prostate cancer: an updated systematic review and meta-analysis', *Nutrients* 10(1) (2018): https://doi.org/10.3390/nu10010040

17. Van Die, M.D., Bone, K.M., Williams, S.G. and Pirotta, M.V., 'Soy and soy isoflavones in prostate cancer: a systematic review and meta-analysis of randomized controlled trials', *British Journal of Urology International* 113(5b) (2014): https://doi.org/10.1111/bju.12435

The Social Conformer

1. *Study of Current and Former Vegetarians and Vegans*, Humane Research Council (2014): https://faunalytics.org/wp-content/uploads/2015/06/Faunalytics_Current-Former-Vegetarians_Full-Report.pdf

2. Orbach, S., 'Britain's obesity strategy ignores the science: dieting doesn't work', *Guardian* (28 July 2020): https://www.theguardian.com/commentisfree/2020/jul/28/britain-obesity-strategy-ignore-science-dieting-calories-stigmatising-fat

The 'What-Abouter'

1. 'Wild bees and not honeybees the main pollinators of UK crops', University of Reading (23 May 2011): https://www.reading.ac.uk/news-archive/press-releases/pr367212.html

2. 'Who are our pollinators?', Department of Agriculture, Enviroment and Rural Affairs: https://www.daera-ni.gov.uk/articles/pollinators-0#:~:text=Who%20are%20our%20pollinators%3F,and%2077%20solitary%20bee%20species

3. Waters, C.M. and Bassler, B.L., 'Quorum sensing: cell-to-cell communication in bacteria', *Annual Review of Cell and Developmental Biology* 21 (2005): https://doi.org/10.1146/annurev.cellbio.21.012704.131001

4. 'Ted Nugent "Vegans kill more animals than non vegans"', YouTube: https://www.youtube.com/watch?v=j0GMSzzYyIA

5. Davis, S.L., 'The least harm principle may require that humans consume a diet containing large herbivores, not a vegan diet', *Journal of Agricultural and Environmental Ethics* 16(4) (2003): https://doi.org/10.1023/A:1025638030686

6. Matheny, G., 'Least harm: a defense of vegetarianism from Steven Davis's omnivorous proposal', *Journal of Agricultural and Environmental Ethics* 16(5) (2003): https://doi.org/10.1023/A:1026354906892

7. Cavia, R., Gómez Villafañe, I.E., Alejandro Cittadino, E., Bilenca, D.N., Miño, M.H. and Busch, M., 'Effects of cereal harvest on abundance and spatial distribution of the rodent *Akodon azarae* in central Argentina', *Agriculture, Ecosystems & Environment* 107(1) (2005): https://doi.org/10.1016/j.agee.2004.09.011

8. Archer, M., 'Ordering the vegetarian meal? There's more animal blood on your hands', The Conversation (15 December 2011): https://theconversation.com/ordering-the-vegetarian-meal-theres-more-animal-blood-on-your-hands-4659

9. *Australia: Grain and Feed Annual*, USDA Foreign Agricultural Service (2019): https://apps.fas.usda.gov/newgainapi/api/report/downloadreportbyfilename?filename=Grain%20and%20Feed%20Annual_Canberra_Australia_5-9-2019.pdf

10. Fischer, B. and Lamey, A., 'Field deaths in plant agriculture', *Journal of Agricultural and Environmental Ethics* 31 (2018): https://doi.org/10.1007/s10806 018 9733 8

11. Ibid.

The Compromiser

1. van Huis, A. and Oonincx, D.G.A.B., 'The environmental sustainability of insects as food and feed. A review', *Agronomy for Sustainable Development* 37 (2017): https://doi.org/10.1007/s13593-017-0452-8

2. Berggren, Å., Jansson, A. and Low, M., 'Approaching ecological sustainability in the emerging insects-as-food industry', *Trends in Ecology and Evolution* 34(2) (2019): https://doi.org/10.1016/j.tree.2018.11.005

3. Alexander, P., Brown, C., Arneth, A., Finnigan, J., Rounsevell, M.D.A., 'Human appropriation of land for food: the role of diet', *Global Environmental Change* 41 (2016): https://doi.org/10.1016/j.gloenvcha.2016.09.005

4. Van Huis, A., Van Itterbeeck, J., Klunder, H., Mertens, E., Halloran, A., Muir, G. and Vantomme, P., *Edible Insects: Future Prospects for Food and Feed Security*, Food and Agruculture Organization of the United Nations (2013): https://www.fao.org/3/i3253e/i3253e.pdf

5. Payne, C., 'Entomophagy: how giving up meat and eating bugs can help save the planet', *Independent* (21 March 2018): https://www.independent.co.uk/news/long_reads/entomophagy-eat-insects-food-diet-save-planet-meat-cattle-deforestation-a8259991.html

6. Alexander, P., Brown, C., Arneth, A., Dias, C., Finnigan, J., Moran, D. and Rounsevell, M.D.A., 'Could consumption of insects, cultured meat or imitation meat reduce global agricultural land use?', *Global Food Security* 15 (2017): https://doi.org/10.1016/j.gfs.2017.04.001

7. Gerretsen, I., 'A neglected protein-rich "superfood"', BBC Future (12 April 2021): https://www.bbc.com/future/article/20210420-the-protein-rich-superfood-most-europeans-wont-eat

8. Lundy, M.E. and Parrella, M.P., 'Crickets are not a free lunch: protein capture from scalable organic side-streams via high-density populations of *Acheta domesticus*', *PLoS ONE* 10(4) (2015): https://doi.org/10.1371/journal.pone.0118785

9. Pitman, J.L., DasGupta, S., Krashes, M.J., Leung, B., Perrat, P.N. and Waddell, S., 'There are many ways to train a fly', *Fly* 3(1) (2009): https://doi.org/10.4161/fly.3.1.7726

10. Nieder, A., 'Honey bees zero in on the empty set', *Science* 360(6393) (2018): https://www.science.org/doi/10.1126/science.aat8958

11. Khuong, T.M., Wang Q.P., Manion, J., Oyston, L.J., Lau, M.T., Towler, H., Lin, Y.Q. and Neely, G.G., 'Nerve injury drives a heightened state of vigilance and neuropathic sensitization in *Drosphila*', *Science Advances* 5(7) (2019): https://doi.org/10.1126/sciadv.aaw4099

12. Zabala, N.A. and Gómez, M.A., 'Morphine analgesia, tolerance and addiction in the cricket *pteronemobius* sp. (orthoptera, insecta)', *Pharmocology Biochemistry and Behavior* 40(4) (1991): https://doi.org/10.1016/0091-3057(91)90102-8

13. Rowe, A., *Insects Raised for Food and Feed – Global Scale, Practices, and Policy*, Rethink Priorities (2020): https://rethinkpriorities.org/publications/insects-raised-for-food-and-feed

14. Node, R., 'The cricket dilemma', Medium (2 April 2019): https://medium.com/bowel-movements/the-cricket-dilemma-feb0a5b65ae2

15. Perry, C.J. and Baciadonna, L., 'Studying emotion in invertebrates: what has been done, what can be measured and what they can provide', *Journal of Experimental Biology* 220(21) (2017): https://doi.org/10.1242/jeb.151308

16. Dyer, A.G., Neumeyer, C. and Chittka, L., 'Honeybee (*Apis mellifera*) vision can discriminate between and recognise images of human faces', *The Journal Experimental Biology* 208(24) (2005): https://doi.org/10.1242/jeb.01929

17. Chittka, L., 'Bee cognition', *Current Biology* 27 (2017): https://www.cell.com/current-biology/pdf/S0960-9822(17)31017-5.pdf

18. Gibbons, M., Versace, E., Crump, A., Baran B. and Chittka, L., 'Motivational trade-offs and modulation of nociception in bumblebees', *Psychological and Cognitive Sciences* 119(31) (2022): https://doi.org/10.1073/pnas.2205821119

19. Best, S., 'Bumblebees CAN feel pain: study shows insects can suppress their withdrawal reflexes in exchange for a sweet treat – suggesting they experience discomfort and "should be included in animal welfare laws"', *Daily Mail*

(27 July 2022): https://www.dailymail.co.uk/sciencetech/article-11053185/Bumblebees-feel-pain-included-animal-welfare-laws-study-claims.html

20.   'Think of honeybees as "livestock" not wildlife, argue experts', University of Cambridge (25 January 2018): https://www.cam.ac.uk/research/news/think-of-honeybees-as-livestock-not-wildlife-argue-experts

21.   Valido, A., Rodríguez-Rodríguez, M.C. and Jordano, P., 'Honeybees disrupt the structure and functionality of plant-pollinator networks', *Scientific Reports* 9 (2019): https://doi.org/10.1038/s41598-019-41271-5

22.   Thomson, D.M., 'Local bumblebee decline linked to recovery of honey bees, drought effects on floral resources', *Ecology Letters* 19(10) (2016): https://doi.org/10.1111/ele.12659

23.   Garibaldi, L.A., Steffan-Dewenter, I., Winfree, R., Aizen, M.A., Bommarco, R., Cunningham, S.A., Kremen, C. et al., 'Wild pollinators enhance fruit set of crops regardless of honey bee abundance', *Science* 339(6127) (2013): https://www.science.org/doi/10.1126/science.1230200

24.   Turner, A., '"Honeybees are voracious": is it time to put the brakes on the boom in beekeeping?', *Guardian* (24 July 2021): https://www.theguardian.com/environment/2021/jul/24/this-only-saves-honeybees-the-trouble-with-britains-beekeeping-boom-aoe

25.   Wijsman, J.W.M., Troost, K., Fang, J. and Roncarati, A., 'Global production of marine bivalves. Trends and challenges', in Smaal, A., Ferreira, J., Grant, J., Petersen, J. and Strand, Ø. (eds), *Goods and Services of Marine Bivalves* (Springer, 2019): https://doi.org/10.1007/978-3-319-96776-9_2

26.   Selbach, C., Marchant, L. and Mouritsen, K.N., 'Mussel memory: can bivalves learn to fear parasites?', *Royal Society Open Science* 9(1) (2022): https://doi.org/ 10.1098/rsos.211774

27.   Hubert, J., Booms, E., Witbaard, R. and Slabbekoorn, H., 'Responsiveness and habituation to repeated sound exposures and pulse trains in blue mussels', *Journal of Experimental Marine Biology and Ecology* 547 (2022): https://doi.org/10.1016/j.jembe.2021.151668

28.   Cadet, P., Zhu, W., Mantione K.J., Baggerman, G. and Stefano, G.B., 'Cold stress alters Mytilus edulis pedal ganglia expression of mu opiate receptor transcripts determined by real-time RT-PCR and morphine levels', *Brain Research. Molecular Brain Research* 99(1) (2002): https://doi.org/10.1016/s0169-328x(01)00342-4

29.   Sonetti, D., Mola, L., Casares, F., Bianchi, E., Guarna, M. and Stefano, G.B., 'Endogenous morphine levels increase in molluscan neural and immune tissues after physical trauma', *Brain Research*, 835(2) (1999): https://doi.org/ 10.1016/s0006-8993(99)01534-6

30.   *RSPCA Welfare Standards for Pigs*, RSPCA (2016): https://science.rspca.org.uk/documents/1494935/9042554/RSPCA+welfare+standards+for+pigs+%28PDF+5.72MB%29.pdf/8b2d5794-9a10-cd1f-f27d-e3642c0c1945?t=1557668440116

31.   *RSPCA Welfare Standards for Dairy Cattle*, RSPCA (2018): https://science.rspca.org.uk/documents/1494935/9042554/1306_DairyCattle_Standards_2022_v2.pdf/ad895c3e-6bb8-fe7a-87a4-b16f10778e1c?t=1672833219745

32.   'MPs ditch plans to immediately ban farrowing crates', FarmingUK (25 November 2021): https://www.farminguk.com/news/mps-ditch-plans-to-immediately-ban-farrowing-crates_59404.html

The Anti-Woke Warrior

1. Milman, O., 'Bill Gates backs new startup aiming to reduce emissions from cow burps', *Guardian* (24 January 2023): https://www.theguardian.com/us-news/2023/jan/24/bill-gates-startup-cow-burps-methane-emissions

2. Bill & Melinda Gates Foundation: https://www.gatesfoundation.org/about/committed-grants/2023/03/inv-050574

3. Neslen, A., ' "The anti-livestock people are a pest": how UN food body played down role of farming in climate change', *Guardian* (20 October 2023): https://www.theguardian.com/environment/2023/oct/20/the-anti-livestock-people-are-a-pest-how-un-fao-played-down-role-of-farming-in-climate-change

4. Neslen, A., 'We need power to prescribe climate policy, IPCC scientists say', *Guardian* (7 December 2023): https://www.theguardian.com/environment/2023/dec/07/we-need-power-to-prescribe-climate-policy-ipcc-scientists-say

5. Horton, H., 'Sunak stands with net zero and climate conspiracy group at farming protest', *Guardian* (25 February 2024): https://www.theguardian.com/politics/2024/feb/25/sunak-stands-with-net-zero-and-climate-conspiracy-group-at-farming-protest

6. Ibid.

7. 'Truth, lies and culture wars: social listening analysis of meat and dairy persuasion narratives', Changing Markets Foundation (2023) : https://changingmarkets.org/wp-content/uploads/2024/01/Truth-lies-and-culture-wars-final.pdf

8. Chan, E.Y. and Zlatevska, N., 'Jerkies, tacos, and burgers: subjective socio-economic status and meat preference', *Appetite* 132 (2019): https://doi.org/10.1016/j.appet.2018.08.027

9. Ibid.

The Well-Intentioned Leftist

1. Yazbeck, J., 'The problem with white veganism', Medium (1 November 2018): https://julianayaz.medium.com/the-problem-with-white-veganism-f86c0341e2a2

2. Butler, P., 'Nearly half of BAME UK households are living in poverty', *Guardian* (1 July 2020): https://www.theguardian.com/society/2020/jul/01/nearly-half-of-bame-uk-households-are-living-in-poverty

3. Funk, C. and Kennedy, B., *The New Food Fights: U.S. Public Divides Over Food Science*, Pew Research Center (2016): https://www.pewresearch.org/internet/wp-content/uploads/sites/9/2016/11/PS_2016.12.01_Food-Science_FINAL.pdf

4. Reiley, L., 'The fastest-growing vegan demographic is African Americans. Wu-Tang Clan and other hip-hop acts paved the way', *Washington Post* (24 January 2020): https://www.washingtonpost.com/business/2020/01/24/fastest-growing-vegan-demographic-is-african-americans-wu-tang-clan-other-hip-hop-acts-paved-way/

5. Springmann, M., Clark, M.A., Rayner, M., Scarborough, P. and Webb, P., 'The global and regional costs of healthy and sustainable dietary patterns: a modelling study', *The Lancet* 5(11) (2021): https://doi.org/10.1016/S2542-5196(21)00251-5

6. Wasley, A. and Heal, A., 'Revealed: shocking safety record of UK meat plants', Bureau of Investigative Journalism (20 July 2018): https://www.thebureauinvestigates.com/stories/2018-07-29/uk-meat-plant-injuries

7. Fitzgerald, A.J., Kalof, L. and Dietz, T., 'Slaughterhouses and increased crime rates: an empirical analysis of the spillover from "The Jungle" into the surrounding community', *Organization and Environment* 22(2) (2009): https://doi.org/10.1177/1086026609338164

8. McDonald, G.G., Costello, C., Bone, J., Cabral, R.B., Farabee, V., Hochberg, T., Kroodsma, D., Mangin, T., Meng, K.C. and Zahn, O., 'Satellites can reveal global extent of forced labor in the world's fishing fleet', *Proceedings of the National Academy of Sciences* 118(3) (2021): https://doi.org/10.1073/pnas.2016238117

9. Mason, M., McDowell, R., Htusan, E. and Mendoza, M., 'Shrimp sold by global supermarkets is peeled by slave labourers in Thailand', *Guardian* (14 December 2015): https://www.theguardian.com/global-development/2015/dec/14/shrimp-sold-by-global-supermarkets-is-peeled-by-slave-labourers-in-thailand

10. Domingo, N.G.G., Balasubramanian, S., Thakrar, S.K., Hill, J.D. et al., 'Air quality-related health damages of food', *Proceedings of the National Academy of Sciences* 118(20) (2021): https://doi.org/10.1073/pnas.2013637118

The Misinformed Environmentalist

1. Johnson, R. and Cody, B.A., *California Agricultural Production and Irrigated Water Use*, Congressional Research Service (2015): https://sgp.fas.org/crs/misc/R44093.pdf

2. *About Almonds and Water*, California Almonds, Almond Board of America (2015): https://www.almonds.com/sites/default/files/content/attachments/about_almonds_and_water_-_september_2015_1.pdf

3. Johnson and Cody, *California Agricultural Production and Irrigated Water Use*

4. 'Fodder for debate: US exports water as hay to sustain Asia's livestock', Gro Intelligence (19 July 2017): https://gro-intelligence.com/insights/us-exports-water-as-hay-to-sustain-asia-livestock

5. Matthews, W.A. and Sumner, D.A., *Contributions of the California Dairy Industry to the California Economy in 2018*, University of California, Agricultural Issues Center (2019): https://aic.ucdavis.edu/wp content/uploads/2019/07/CMAB-Economic-Impact-Report_final.pdf

6. Richter, B.D., Bartak, D., Caldwell, P., Davis, K.F., Debaere, P., Hoekstra, A.Y., Li, T., Marston, L., McManamay, R., Mekonnen, M.M., Ruddell, B.L., Rushforth, R.R. and Troy, T.J., 'Water scarcity and fish imperilment driven by beef production', *Nature Sustainability* 3 (2020): https://doi.org/10.1038/s41893-020-0483-z

7. Poore, J. and Nemecek, T., 'Reducing food's environmental impacts through producers and consumers', *Science* 360(6392) (2018): https://www.science.org/doi/10.1126/science.aaq0216

8. Guibourg, C. and Briggs, H., 'Climate change: which vegan milk is best?', BBC News (22 February 2019): https://www.bbc.co.uk/news/science-environment-46654042

9. Poore and Nemecek, 'Reducing food's environmental impacts through producers and consumers'

10. Ajiero, I. and Campbell, D., 'Benchmarking water use in the UK food and drink sector: case study of three water-intensive dairy products', *Water Conservation Science and Engineering* 3 (2018): https://doi.org/10.1007/s41101-017-0036-0

11. Ibid.

12. De Ruiter, H., Macdiarmid, J.I., Matthews, R.B., Kastner, T. and Smith, P., 'Global cropland and greenhouse gas impacts of UK food supply are increasingly located overseas', *Journal of the Royal Society Interface* 13(114) (2016): https://doi.org/10.1098/rsif.2015.1001

13. Marshall, C., and Prior, M., 'Livestock farming polluted rivers 300 times in one year', BBC News (16 December 2022): https://www.bbc.co.uk/news/science-environment-63961659

14. Cutcher, N., 'Two-thirds of cattle farms in north Devon cause river pollution', *Guardian* (25 October 2022): https://www.theguardian.com/environment/2022/oct/25/two-thirds-of-cattle-farms-in-north-devon-cause-river-pollution

15. Marshall and Prior, 'Livestock farming polluted rivers 300 times in one year'

16. Vanham, D., Comero, S., Gawlik, B.M. and Bidoglio, G., 'The water footprint of different diets within European sub-national geographical entities', *Nature Sustainability* 1 (2018): https://doi.org/10.1038/s41893-018-0133-x

17. Ritchie, H. and Roser, M., 'Soy', Our World in Data (2021): https://ourworldindata.org/soy

18. Ritchie, H., 'Dairy vs. plant-based milk: what are the environmental impacts?', Our World in Data (19 January 2022): https://ourworldindata.org/environmental-impact-milks

19. Ritchie and Roser, 'Soy'

20. 'Nowhere to hide: how the fashion industry is linked to Amazon Rainforest destruction', Stand.earth (29 November 2021): https://stand.earth/resources/nowhere-to-hide-how-the-fashion-industry-is-linked-to-amazon-rainforest-destruction/

21. Fraanje, W., 'Soy in the UK: what are its uses?' Table (20 February 2020): https://www.tabledebates.org/blog/soy-uk-what-are-its-uses

22. *Risky Business*, WWF and RSPB (2017): https://www.wwf.org.uk/sites/default/files/2017-10/WWF%20and%20RSPB%20-%20Risky%20Business%20Report%20-%20October%202017.pdf

23. Fraanje, 'Soy in the UK: what are its uses?'

24. Ritchie, H. and Roser, M., 'Palm oil', Our World in Data (December 2020): https://ourworldindata.org/palm-oil

25. Ibid.

26. Ibid.

27. Ibid.

28. Ibid.

29. Ibid.

30. 'How can we use palm oil without destroying tropical forests?' in Ritchie and Roser, 'Palm oil': https://ourworldindata.org/palm-oil#how-can-we-use-palm-oil-without-destroying-tropical-forests

31. *Sustainability of Liquid Biofuels*, Royal Academy of Engineering (2017): https://raeng.org.uk/media/pk1k5mie/raeng-biofuels-report-1-1.pdf

32. *Mapping and Understanding the UK Palm Oil Supply Chain*, Department for Environment, Food and Rural Affairs (2011): https://www.plant-talk.org/documents/UKpalmoil.pdf

33. Ibid.

34. Ritchie, H., 'Cutting down forests: what are the drivers of deforestation?', Our World in Data (23 February 2021): https://ourworldindata.org/what-are-drivers-deforestation

35. Machovina, B., Feeley, K.J. and Ripple, W.J., 'Biodiversity conservation: the key is reducing meat consumption', *Science of the Total Environment* 536 (2015): https://doi.org/10.1016/j.scitotenv.2015.07.022

36. Benton, T.G., Bieg, C., Harwatt, H., Pudasaini, R. and Wellesley L., *Food System Impacts on Biodiversity Loss*, Chatham House (2021): https://www.chathamhouse.org/sites/default/files/2021-02/2021-02-03-food-system-biodiversity-loss-benton-et-al_0.pdf

37. Kite-Powell, J., 'Using drip irrigation to make new sustainable growing regions for avocados', *Forbes* (29 March 2022): https://www.forbes.com/sites/jenniferhicks/2022/03/29/using-drip-irrigation-to-make-new-sustainable-growing-regions-for-avocados/

38. Chu, J., 'Watering the world' MIT News (19 April 2017): https://news.mit.edu/2017/design-cuts-costs-energy-drip-irrigation-0420

39. Ritchie, H., 'You want to reduce the carbon footprint of your food? Focus on what you eat, not whether your food is local', Our World in Data (24 January 2020): https://ourworldindata.org/food-choice-vs-eating-local

40. Clune, S., Crossin, E. and Verghese, K., 'Systematic review of greenhouse gas emissions for different fresh food categories', *Journal of Cleaner Production* 140(2) (2017): https://doi.org/10.1016/j.jclepro.2016.04.082

41. Finney, C., End of the avocado: why chefs are ditching the unsustainable fruit', *Guardian* (1 November 2021): https://www.theguardian.com/food/2021/nov/01/end-of-the-avocado-why-chefs-ditching-the-unsustainable-fruit

42. Poore, and Nemecek, 'Reducing food's environmental impacts through producers and consumers'

43. De Ruiter, H., Macdiarmid, J.I., Matthews, R.B., Kastner, T., Lynd, L.R. and Smith, P., 'Total global agricultural land footprint associated with UK food supply 1986–2011', *Global Environmental Change* 43 (2017): https://doi.org/10.1016/j.gloenvcha.2017.01.007

44. Harwatt, H. and Hayek, M.N., *Eating Away at Climate Change: Repurposing UK Agricultural Land to Meet Climate Goals*, Harvard Law School (2019): https://animal.law.harvard.edu/wp-content/uploads/Eating-Away-at-Climate-Change-with-Negative-Emissions%E2%80%93%E2%80%93Harwatt-Hayek.pdf

45. 'GM in animal feed', Food Standards Agency: https://www.food.gov.uk/business-guidance/gm-in-animal-feed

46. Merrill, D. and Leatherby, L., 'Here's how America uses its land', Bloomberg (31 July 2018): https://www.bloomberg.com/graphics/2018-us-land-use/?in_source=embedded-checkout-banner

47. Harwatt and Hayek, *Eating Away at Climate Change*

48. Monbiot, G., 'Interview: George Monbiot: "On a vegan planet, Britain could feed 200 million people"', *Guardian* (13 May 2022): https://www.theguardian.com/culture/2022/may/13/george-monbiot-vegan-planet-britain-farming-fuel-plant-based-food

49. 'Government urged to back British food at home and abroad', NFU (15 September 2021): https://www.nfuonline.com/updates-and-information/government-urged-to-back-british-food-at-home-and-abroad/

50. Harwatt and Hayek, *Eating Away at Climate Change*

51. *Bullish Prospects: A Vision for the Beef Industry*, NFU (2012): https://www.nfuonline.com/archive?treeid=15630

52. Poore and Nemecek, 'Reducing food's environmental impacts through producers and consumers'

53. Ranganathan, J., Waite, R., Searchinger, T. and Hanson, C., *How to Sustainably Feed 10 Billion People by 2050, in 21 Charts*, World Resources Institute (2018): https://www.wri.org/insights/how-sustainably-feed-10-billion-people-2050-21-charts

54. Harwatt and Hayek, *Eating Away at Climate Change*

55. 'Share of cereals allocated to food, animal feed or fuel, 2020', Our World in Data: https://ourworldindata.org/grapher/cereal-allocation-by-country

56. Palmer, R.C. and Smith, R.P., 'Soil structural degradation in SW England and its impact on surface-water runoff generation', *Soil Use and Management* 29(4) (2013): https://doi.org/10.1111/sum.12068

57. *Farming Statistics – Land Use, Livestock Populations and Agricultural Workforce at 1 June 2020 – England*, Department for Environment, Food and Rural Affairs (2020): https://assets.publishing.service.gov.uk/government/uploads/system/uploads/attachment_data/file/928397/structure-landuse-june20-eng-22oct20.pdf

58. 'Lowland grassland', NatureScot: https://www.nature.scot/landscapes-and-habitats/habitat-types/farmland-and-croftland/lowland-grassland

59. Harwatt and Hayek, *Eating Away at Climate Change*

60. Merrill and Leatherby, 'Here's how America uses its land'

61. Carter, N. and Mehta, T., 'Another failed attempt to greenwash beef', Sentient Media (14 January 2021): https://sentientmedia.org/another-failed-attempt-to-greenwash-beef/

62. Ibid.

63. *Land Use: Policies for a Net Zero UK*, Committee on Climate Change (2020): https://www.theccc.org.uk/publication/land-use-policies-for-a-net-zero-uk/

64. Pieper, M., Michalke, A. and Gaugler, T., 'Calculation of external climate costs for food highlights inadequate pricing of animal products', *Nature Communications* 11(1) (2020): https://doi.org/10.1038/s41467-020-19474-6

65. Roque, B.M. et al., 'Red seaweed *(Asparagopsis taxiformis)* supplementation reduces enteric methane by over 80 percent in beef steers', *PLoS One* 16(3) (2021): https://doi.org/10.1371/journal.pone.0247820

66. Roque, B.M. et al., 'Inclusion of *Asparagopsis armata* in lactating dairy cows' diet reduces enteric methane emission by over 50 percent', *Journal of Cleaner Production* 234 (2019): https://doi.org/10.1016/j.jclepro.2019.06.193

67. Kinley, R.D. et al., 'Mitigating the carbon footprint and improving productivity of ruminant livestock agriculture using a red seaweed', *Journal of Cleaner Production* 259 (2020): https://doi.org/10.1016/j.jclepro.2020.120836

68. 'Effect of Asparagopsis extract in a canola oil carrier for long-fed Wagyu cattle', Meat and Livestock Australia (2023): https://www.mla.com.au/research-and-development/reports/2023/p.psh.1353---effect-of-asparagopsis-extract-in-a-canola-oil-carrier-for-long-fed-wagyu-cattle/

69. Camer-Pesci, B. et al., 'Opportunities of *Asparagopsis* sp. cultivation to reduce methanogenesis in ruminants: a critical review', *Algal Research* 76: https://doi.org/10.1016/j.algal.2023.103308

70. Ritchie, 'You want to reduce the carbon footprint of your food? Focus on what you eat, not whether your food is local'

71. Poore, and Nemecek, 'Reducing food's environmental impacts through producers and consumers'

72. Sandström, V., Valin, H., Krisztin, T., Havlík, P., Herrero, M. and Kastner, T., 'The role of trade in the greenhouse gas footprints of EU diets', *Global Food Security* 19 (2018): https://doi.org/10.1016/j.gfs.2018.08.007

73. Scarborough, P., Clark, M., Cobiac, L., Papier, K., Knuppel, A., Lynch, J., Harrington, R., Key, T. and Springmann, M., 'Vegans, vegetarians, fish-eaters and meat-eaters in the UK show discrepant environmental impacts', *Nature Food* 4 (2023): https://doi.org/10.1038/s43016-023-00795-w

74. Weber, C.L. and Matthews, H.S., 'Food-miles and the relative climate impacts of food choices in the United States', *Environmental Science and Technology* 42(10) (2008): https://pubs.acs.org/doi/full/10.1021/es702969f

75. Carlsson-Kanyama, A., Ekström, M.P. and Shanahan, H., 'Food and life cycle energy inputs: consequences of diet and ways to increase efficiency', *Ecological Economics* 44(2–3) (2003): https://doi.org/10.1016/S0921-8009(02)00261-6

76. Hospido, A., Milà i Canals, L., McLaren, S., Truninger, M., Edwards-Jones, G. and Clift, R., 'The role of seasonality in lettuce consumption: a case study of environmental and social aspects', *International Journal of Life Cycle Assessment* 14 (2009): https://doi.org/10.1007/s11367-009-0091-7

77. Poore and Nemecek, 'Reducing food's environmental impacts through producers and consumers'

78. 'AHDB to host mission to the US to grow red meat exports', Food Voices (30 January 2023): https://foodvoices.co.uk/2023/01/ahdb-to-host-mission-to-the-us-to-grow-red-meat-exports/

79. 'International Trade Secretary visits Yorkshire farm to celebrate Great British Beef Week', NFU (28 April 2021): https://www.nfuonline.com/updates-and-information/international-trade-secretary-visits-yorkshire-farm-to-celebrate-great-british-beef-week-1/

80. Holland, L., 'Veganuary is "just a gimmick and causing British industry to the wall", farmers warn', Sky News (6 January 2021): https://news.sky.com/story/veganuary-is-just-a-gimmick-and-casting-british-industry-to-the-wall-farmers-warn-12174067

81. Aubrey, A., 'Your love of quinoa is good news for Andean farmers', The Salt (17 July 2013): https://www.npr.org/sections/thesalt/2013/07/16/202737139/is-our-love-of-quinoa-hurting-or-helping-farmers-who-grow-it

82. Ibid.

83. Bellemare, M.F., Fajardo-Gonzalez, J. and Gitter, S.R., *Foods and Fads: The Welfare Impacts of Rising Quinoa Prices in Peru*, Towson University Department of Economics (2016): https://webapps.towson.edu/cbe/economics/workingpapers/2016-06.pdf

84. Cherfas, J., 'Your quinoa habit really did help Peru's poor. But there's trouble ahead', The Salt (31 March 2016): https://www.npr.org/sections/thesalt/2016/03/31/472453674/your-quinoa-habit-really-did-help-perus-poor-but-theres-trouble-ahead

85. Stevens, A.W., 'Quinoa quandary: cultural tastes and nutrition in Peru', *Food Policy* 71 (2017): https://doi.org/10.1016/j.foodpol.2017.08.003

86. Cherfas, J., 'Your quinoa habit really did help Peru's poor. But there's trouble ahead'

87. Ritchie, H., 'Food waste is responsible for 6% of global greenhouse emissions', Our World in Data (18 March 2020): https://ourworldindata.org/food-waste-emissions

88. Poore and Nemecek, 'Reducing food's environmental impacts through producers and consumers'

89. Ritchie, H., 'How much of global greenhouse emissions come from food?', Our World in Data (18 March 2021): https://ourworldindata.org/greenhouse-gas-emissions-food

The Animal Farmer

1. McCullough, C., 'Tears as young boy says goodbye to slaughter-bound lamb he reared', *Irish Examiner* (20 August 2018): https://www.irishexaminer.com/lifestyle/arid-30863308.html

2. Driver, A., 'Do not betray the pig sector in future US–UK trade deal – NPA', National Pig Association (5 June 2019): http://www.npa-uk.org.uk/Do_not_betray_the_pig_sector_in_future_US-UK_trade_deal-NPA.html

3. 'How much cropland has the world spared due to increases in crop yields?, 2021', Our World in Data: https://ourworldindata.org/grapher/land-sparing-by-crop

4. Ritchie, H., 'Excess fertilizer: which countries cause environmental damage by overapplying fertilizers?', Our World in Data: https://ourworldindata.org/excess-fertilizer

5. Cui, Z., Zhang, H., Chen, X., Zhang, C., Ma, W. et al., 'Pursuing sustainable productivity with millions of smallholder farmers' *Nature* 555 (2018): https://doi.org/10.1038/nature25785

6. Marshall, C. and Prior, M., 'Livestock farming polluted rivers 300 times in one year', BBC News (16 December 2022): https://www.bbc.co.uk/news/science-environment-63961659

7. Carter, N. and Urbancic, N., 'Seeing stars: the new metric that could allow the meat and dairy industry to avoid climate action', Changing Markets Foundation (2023): https://changingmarkets.org/wp-content/uploads/2023/11/Seeing-stars-report.pdf

8. Ibid.

9. McKie, R., 'Nearly half of Britain's biodiversity has gone since industrial revolution', *Guardian* (10 October 2021): https://www.theguardian.com/environment/2021/oct/10/nearly-half-of-britains-biodiversity-has-gone-since-industrial-revolution

10. Filazzola, A., Brown, C., Dettlaff, M.A., Batbaatar, A., Grenke, J., Bao, T., Peetoom Heida, I. and Cahill Jr, J.F., 'The effects of livestock grazing on biodiversity are multi-trophic: a meta-analysis', *Ecology Letters* 23(8) (2020): https://doi.org/10.1111/ele.13527

11. Ibid.

12. Ritchie, H and Roser, M., 'Greenhouse gas emissions', Our World in Data: https://ourworldindata.org/greenhouse-gas-emissions

13. Hainer, M., 'Hummus is so popular, tobacco farmers switch to chickpeas', Today (1 May 2013): https://www.today.com/food/hummus-so-popular-tobacco-farmers-switch-chickpeas-6C9692848

14. *Thriving Beyond the Protein Transition: Farmer and Crofter Receptiveness to Stockfree Land Management*, Stockfree Farming (2022): https://drive.google.com/file/d/16hI6ZwILOO-9P5gfJwIIutaQIUo5bAJk/view

The Pseudoscientist

1. *2021 UK Greenhouse Gas Emissions, Provisional Figures*, Department for Business, Energy and Industrial Strategy (2022): https://assets.publishing. service.gov.uk/government/uploads/system/uploads/attachment_data/ file/1064923/2021-provisional-emissions-statistics-report.pdf

2. White, R.R. and Hall, M.B., 'Nutritional and greenhouse gas impacts of removing animals from US agriculture', *Proceedings of the National Academy of Sciences of the United States of America* 114(48) (2017): https://doi. org/10.1073/pnas.1707322114

3. Elgin, B., 'Beef industry tries to erase its emissions with fuzzy methane math', Bloomberg (19 October 2021): https://www.bloomberg.com/news/features/ 2021-10-19/beef-industry-falsely-claims-low-cow-carbon-footprint

4. Clark, M.A., Domingo, N.G.G., Colgan, K., Thakrar, S.K., Tilman, D., Lynch, J., Azevedo, I.L. and Hill J.D., 'Global food system emissions could preclude achieving the 1.5°C and 2°C climate change targets', *Science* 370(6517) (2020): https://www.science.org/doi/10.1126/science.aba7357

5. Ivanovich, C.C., Sun, T., Gordon, D.R. and Ocko, I.B., 'Future warming from global food consumption', *Nature Climate Change* 13 (2023): https:// doi.org/10.1038/s41558-023-01605-8

6. Sans, P. and Combris, P., 'World meat consumption patterns: an overview of the last fifty years (1961–2011)', *Meat Science* 109 (2015): https://doi. org/10.1016/j.meatsci.2015.05.012

7. Emery, I., 'Without animals, US farmers would reduce feed crop production', *Proceedings of the National Academy of Sciences of the United States of America* 115(8) (2018): https://doi.org/10.1073/pnas.1720760115

8. 'Is soy production driving deforestation?', in Ritchie, H. and Roser, M., 'Soy', Our World in Data: https://ourworldindata.org/soy#is-soy-production- driving-deforestation

9. Myhre, G., Shindell, D., Bréon, F.M., Collins, W., Fuglestvedt, J., Huang, J. et al, 'Anthropogenic and Natural Radiative Forcing', in Stocker, T.F., Qin, D., Plattner, G.K., Tignor, M., Allen, S.K., Boschung, J., et al. (eds), *Climate Change 2013: The Physical Science Basis. Contribution of Working Group I to the Fifth Assessment Report of the Intergovernmental Panel on Climate Change* (Cambridge University Press, 2013): https://www.ipcc.ch/site/ assets/uploads/2018/02/WG1AR5_Chapter08_FINAL.pdf

10. Ibid.

11. Elgin, 'Beef industry tries to erase its emissions with fuzzy methane math'

12. Ibid.

13. Ibid.

14. 'Global assessment: urgent steps must be taken to reduce methane emissions this decade', UN Environment Programme (2021): https://www.unep.org/ news-and-stories/press-release/global-assessment-urgent-steps-must-be- taken-reduce-methane

15. Brown, A., 'Methane emissions: Australian cattle industry suggests shift from net zero target to "climate neutral" approach', *Guardian* (7 May 2024): https://www.theguardian.com/australia-news/article/2024/may/08/ methane-emissions-australian-cattle-industry-suggests-shift-from-net-zero- target-to-climate-neutral-approach

16. Carter, N. and Urbancic, N., 'Seeing stars: the new metric that could allow the meat and dairy industry to avoid climate action', Changing Markets

Foundation (2023): https://changingmarkets.org/wp-content/uploads/2023/11/Seeing-stars-report.pdf

17. 'Latest emissions data', Environmental Protection Agency: https://www.epa.ie/our-services/monitoring--assessment/climate-change/ghg/latest-emissions-data/

18. Carter and Urbancic, 'Seeing stars': https://changingmarkets.org/wp-content/uploads/2023/11/Seeing-stars-report.pdf

19. 'NFU calls for new methane metric to be used in GHG calculations', NFU (31 May 2023): https://www.nfuonline.com/updates-and-information/nfu-calls-for-new-methane-metric-to-be-used-in-ghg-calculations/#:~:text=The%20NFU%20supports%20adoption%20of,NFU's%20Livestock%20and%20Dairy%20boards

20. Sterman, J.D., Siegel, L. and Rooney-Varga, J.N., 'Does replacing coal with wood lower CO_2 emissions? Dynamic lifecycle analysis of wood bioenergy', *Environmental Research Letters* 13 (2018): https://iopscience.iop.org/article/10.1088/1748-9326/aaa512/meta

21. 'Written evidence submitted by Greenpeace UK', Greenpeace (November 2021): https://committees.parliament.uk/writtenevidence/40618/pdf/

22. Crowley, J., 'Drax: UK power station still burning rare forest wood', BBC News (28 February 2024): https://www.bbc.co.uk/news/science-environment-68381160

23. Harrison, T. and Fox, H., 'Biomass plant is UK's top emitter', Ember (31 July 2023): https://ember-climate.org/insights/research/drax-co2-emissions-biomass/

24. 'Drax and NFU to partner boost UK energy crop market', NFU (20 September 2021): https://www.nfuonline.com/media-centre/releases/drax-and-nfu-partner-to-boost-uk-energy-crop-market/

25. Carrington, D., 'No need to cut beef to tackle climate crisis, say farmers', *Guardian* (10 September 2019): https://www.theguardian.com/environment/2019/sep/10/no-need-to-cut-beef-to-tackle-climate-crisis-say-farmers

26. Drax and NFU to partner boost UK energy crop market': https://www.nfuonline.com/media-centre/releases/drax-and-nfu-partner-to-boost-uk-energy-crop-market/

27. 'The plan', National Food Strategy (2021): https://www.nationalfoodstrategy.org/wp-content/uploads/2021/07/National-Food-Strategy-Chapter-16.pdf

28. Horton, H. and Harvey, F., 'England must reduce meat intake to avoid climate breakdown, says food tsar', *Guardian* (16 August 2022): https://www.theguardian.com/environment/2022/aug/16/england-must-reduce-meat-dairy-intake-says-henry-dimbleby

29. Cummings, K.M., Morley, C.P. and Hyland, A., 'Failed promises of the cigarette industry and its effect on consumer misperceptions about the health risks of smoking', *Tobacco Control* 11 (2002): https://tobaccocontrol.bmj.com/content/11/suppl_1/i110

30. 'Jordan Peterson on the food pyramid', YouTube: https://www.youtube.com/watch?v=SVoqFqlBn1k&ab_channel=JPEducation

31. 'Jordan Peterson reacts to Oxford union vegan debate – Joe Rogan part 1', YouTube: https://www.youtube.com/watch?v=5yKtm09Ng2I&ab_channel=IDWTv

32. Lennerz, B.S., Mey, J.T., Henn, O.H. and Ludwig, D.S., 'Behavioral characteristics and self-reported health status among 2029 adults consuming a "carnivore diet"', *Current Developments in Nutrition* 5(12) (2021): https://www.ncbi.nlm.nih.gov/pmc/articles/PMC8684475/

33. Wright, N., Wilson, L., Smith, M., Duncan, B and McHugh, P., 'The BROAD study: a randomised controlled trial using a whole food plant-based diet in the community for obesity, ischaemic heart disease or diabetes', *Nutrition and Diabetes* 7(3) (2017): https://doi.org/10.1038/nutd.2017.3

34. Müller, A., Zimmermann-Klemd, A.M., Lederer, A.K., Hannibal, L., Kowarschik, S., Huber, R. and Storz, M.A., 'A vegan diet is associated with a significant reduction in dietary acid load: post hoc analysis of a randomized controlled trial in healthy individuals', *International Journal of Environmental Research and Public Health* 18(19) (2021): https://doi.org/10.3390/ijerph18199998

35. Kashyap, A., Mackay, A., Carter, B., Fyfe, C.L., Johnstone, A.M. and Myint, P.K., 'Investigating the effectiveness of very low-calorie diets and low-fat vegan diets on weight and glycemic markers in type 2 diabetes mellitus: a systematic review and meta-analysis', *Nutrients* 14(22) (2022): https://doi.org/10.3390/nu14224870

36. 'Bacon's back on the menu study finds', This Morning: https://www.facebook.com/watch/?v=511629266325367

37. Johnston, B.C., Zeraatkar, D., Ah Han, M., Vernooij, R.W.M., Valli, C., El Dib, R., Marshall, C., Stover, P.J., Fairweather-Taitt, S., et al., 'Unprocessed red meat and processed meat consumption: dietary guideline recommendations from the Nutritional Recommendations (NutriRECS) Consortium', *Annals of Internal Medicine* 117 (2019): https://doi.org/10.7326/M19-1621

38. Erickson, J., Sadeghirad, B., Lytvyn, L., Slavin, J. and Johnston, B.C., 'The scientific basis of guideline recommendations on sugar intake: a systematic review', *Annals of Internal Medicine* 166(4) (2017): https://doi.org/10.7326/M16-2020

39. Jacobs, A., 'A shadowy industry group shapes food policy around the world' *New York Times* (16 September 2019): https://www.nytimes.com/2019/09/16/health/ilsi-food-policy-india-brazil-china.html

40. Parker-Pope, T. and O'Connor, A., 'Scientist who discredited meat guidelines didn't report past food industry ties', *New York Times* (4 October 2019): https://www.nytimes.com/2019/10/04/well/eat/scientist-who-discredited-meat-guidelines-didnt-report-past-food-industry-ties.html

41. Oreskes, N., 'So is it okay to eat more red and processed meat?' *Scientific American* (1 February 2020): https://www.scientificamerican.com/article/so-is-it-okay-to-eat-more-red-and-processed-meat/

42. Vogiatzoglou, A., Refsum, H., Johnston, C., Smith, S.M., Bradley, K.M., de Jager, C., Budge, M.M. and Smith, A.D., 'Vitamin B12 status and rate of brain volume loss in community-dwelling elderly', *Neurology* 71(11) (2008): https://doi.org/10.1212/01.wnl.0000325581.26991.f2

43. 'Vegetable-only diet ups risk for brain shrinkage', Fox News (13 January 2015): https://www.foxnews.com/story/vegetable-only-diet-ups-risk-for-brain-shrinkage

44. Spencer, B., 'Trendy vegan diets could LOWER IQ due to lack of a nutrient that is critical to brain health, leading nutritionist warns' [published in print as 'Momentum of veganism threatening brain health of the UK, expert warns'], *Daily Mail* (30 August 2019): https://www.dailymail.co.uk/health/article-7407269/Momentum-veganism-threatening-brain-health-UK-expert-warns.html

45. McDermott, N., 'Plant-based diets risk "dumbing down" the next generation, nutritionist warns', *The Sun* (30 August 2019): https://nypost.com/2019/08/30/plant-based-diets-risk-dumbing-down-the-next-generation-nutritionist-warns/

46. Derbyshire, E., 'Could we be overlooking a potential choline crisis in the United Kingdom?', *BMJ Nutrition, Prevention & Health* 2(2) (2019): https://doi.org/10.1136/bmjnph-2019-000037

47. Parkinson, C., 'The brain nutrient vegans need to know about', BBC News (30 August 2019): https://www.bbc.co.uk/news/health-49509504

48. Patterson, K.Y., Bhagwat, S.A., Williams, J.R., Howe, J.C., Holden, J.M., Zeisel, S.H., Dacosta, K.A. and Mar, M.H., *USDA Database for the Choline Content of Common Foods, Release Two*, US Department of Agriculture (2008): https://data.nal.usda.gov/system/files/Choln02.pdf

49. Derbyshire, 'Could we be overlooking a potential choline crisis in the United Kingdom?'

50. Mazidi, M. and Kengne, A.P., 'Higher adherence to plant-based diets are associated with lower likelihood of fatty liver', *Clinical Nutrition* 38(4) (2019): https://doi.org/10.1016/j.clnu.2018.08.010

51. Gleckel, J.A., 'Are consumers really confused by plant-based food labels? An empirical study', *Journal of Animal and Environmental Law* (2021): https://ssrn.com/abstract=3727710

52. Dewan, P., 'Sustainability scientist debunks oat vs dairy milk comparison', Better Planet (15 September 2023): https://www.newsweek.com/sustainability-scientist-debunks-oat-dairy-milk-comparison-1827477

53. Ritchie, H., 'Dairy vs. plant-based milk: what are the environmental impacts?', Our World in Data (19 January 2022): https://ourworldindata.org/environmental-impact-milks

54. Ritchie, H., 'What are the carbon opportunity costs of our food?', Our World in Data (19 March 2021): https://ourworldindata.org/carbon-opportunity-costs-food

55. 'A good guide to good carbs: the glycemic index', Harvard Health Publihshing (14 April 2023): https://www.health.harvard.edu/healthbeat/a-good-guide-to-good-carbs-the-glycemic-index

56. 'The lowdown on glycemic index and glycemic load', Harvard Health Publishing (2 August 2023): https://www.health.harvard.edu/diseases-and-conditions/the-lowdown-on-glycemic-index-and-glycemic-load

57. Ibid.

58. Jones, B., 'Is oat milk unhealthy? That's the wrong question', Vox (25 February 2024): https://www.vox.com/future-perfect/24072187/is-oat-milk-bad-for-you-or-healthy-wrong-question

59. Baggini, J., '"Personalising stuff that doesn't matter": the trouble with the Zoe nutrition app', *Guardian* (18 May 2024): https://www.theguardian.com/science/article/2024/may/18/zoe-nutrition-app-diet-tim-spector-wellness-science

60. Coffey, H., 'The rise and fall of oat milk: has the trendiest dairy alternative finally fallen from grace?', *Independent* (5 February 2024): https://www.independent.co.uk/life-style/food-and-drink/features/oat-milk-nutrition-dairy-free-b2489539.html

61. 'WHO report says eating processed meat is carcinogenic: understanding the findings', Harvard T.H. Chan School of Public Health (3 November 2015): https://nutritionsource.hsph.harvard.edu/2015/11/03/report-says-eating-processed-meat-is-carcinogenic-understanding-the-findings/

62. Aune, D. et al., 'Whole grain consumption and risk of cardiovascular disease, cancer, and all cause and cause specific mortality: systematic review and dose-response meta-analysis of prospective studies', *British Medical Journal* 352 (2016): https://doi.org/10.1136/bmj.i2716

63. Ying, T. et al., 'Effects of whole grains on glycemic control: a systematic review and dose-response meta-analysis of prospective cohort studies and randomized controlled trials', *Nutrition Journal* 23 (2024): https://nutritionj.biomedcentral.com/articles/10.1186/s12937-024-00952-2

64. Hou Q. et al., 'The metabolic effects of oats intake in patients with type 2 diabetes: a systematic review and meta-analysis, *Nutrients* 7(12) (2015): https://doi.org/10.3390/nu7125536

65. Qian, F., Liu, G., Hu, F.B., Bhupathiraju, S.N. and Sun, Q., 'Association between plant-based dietary patterns and risk of type 2 diabetes: a systematic review and meta-analysis', *JAMA Internal Medicine* 179(10) (2019): https://doi.org/10.1001/jamainternmed.2019.2195

66. McMacken, M. and Shah, S., 'A plant-based diet for the prevention and treatment of type 2 diabetes', *Journal of Geriatric Cardiology* 14(5) (2017): https://pubmed.ncbi.nlm.nih.gov/28630614/

67. Allan Savory, 'How to fight desertification and reverse climate change', TED Talks: https://www.ted.com/talks/allan_savory_how_to_fight_desertification_and_reverse_climate_change

68. Ketcham, C., 'Allan Savory's holistic management theory falls short on science', *Sierra* (23 Febaury 2017): https://www.sierraclub.org/sierra/2017-2-march-april/feature/allan-savory-says-more-cows-land-will-reverse-climate-change

69. Briske, D.D., Bestelmeyer, B.T., Brown, J.R., Fuhlendorf, S.D. and Polley, H.W., 'The Savory method can not green deserts or reverse climate change', *Rangelands* 35(5) (2013): https://www.ars.usda.gov/ARSUserFiles/4472/RANGELANDS-D-13-00044.pdf

70. Garnett, T., Godde, C., Muller, A., Röös, E., Smith,P., de Boer, I., zu Ermgassen, E., Herrero, M., van Middelaar, C., Schader, C. and van Zanten, H, *Grazed and Confused*, Food Climate Research Network (2017): https://tabledebates.org/node/12335

71. *Agricultural Statistics and Climate Change*, Department for Environment, Food and Rural Affairs (2020): https://assets.publishing.service.gov.uk/government/uploads/system/uploads/attachment_data/file/941991/agriclimate-10edition-08dec20.pdf

The Naturalist

1. Dunn, K., 'Farm confessional: what butchering your animals really feels like', Modern Farmer (15 October 2014): https://modernfarmer.com/2014/10/butchering-animals/

2. *National Survey of Fishing, Hunting, and Wildlife-Associated Recreation*, US Fish and Wildlife Service (2016): https://www.census.gov/content/dam/Census/library/publications/2018/demo/fhw16-nat.pdf

3. 'What to do about deer', Humane Society of the United States: https://www.humanesociety.org/resources/what-do-about-deer

4. Tobias, J., 'The secretive government agency planting "cyanide bombs" across the US', *Guardian* (26 June 2020): https://www.theguardian.com/environment/2020/jun/26/cyanide-bombs-wildlife-services-idaho

5. 'Venison meat yield', Ohio Department of Natural Resources.

6. Hussain, G., 'How many animals are killed for food every day?', Sentient Media (31 August 2022): https://sentientmedia.org/how-many-animals-are-killed-for-food-every-day/

NOTES

The Amateur Nutritionist

1. Mariotti, F. and Gardner C.D., 'Dietary protein and amino acids in vegetarian diets —a review', *Nutrients* 11(11) (2019): https://doi.org/10.3390/nu11112661

2. Rizzo, N.S., Jaceldo-Siegl, K., Sabate, J. and Fraser, G.E., 'Nutrient profiles of vegetarian and nonvegetarian dietary patterns', *Journal of the Academy of Nutrition and Dietetics* 113(12) (2013): https://doi.org/10.1016/j.jand.2013.06.349

3. Hevia-Larraín, V., Gualano, B., Longobardi, I., Gil, S., Fernandes, A.L., Costa, L.A.R., Pereira, R.M.R., Artioli, G.G., Phillips, S.M. and Roschel, H., 'High-protein plant-based diet versus a protein-matched omnivorous diet to support resistance training adaptations: a comparison between habitual vegans and omnivores', *Sports Medicine* 51(6) (2021): https://doi.org/10.1007/s40279-021-01434-9

4. Li, C.Y., Fan, A.P., Ma, W.J., Wu S.L., Li C.L., Chen Y.M. and Zhu, H.L., 'Amount rather than animal vs plant protein intake is associated with skeletal muscle mass in community-dwelling middle-aged and older Chinese adults: results from the Guangzhou nutrition and health study', *Journal of the Academy of Nutrition and Dietetics* 119(9): https://doi.org/10.1016/j.jand.2019.03.010

5. Katz, D.L., Doughty, K.N., Geagan, K., Jenkins, D.A. and Gardner, C.D., 'Perspective: the public health case for modernizing the definition of protein quality', *Advances in Nutrition* 10(5) (2019): https://doi.org/10.1093/advances/nmz023

6. Herreman, L., Nommensen, P., Pennings, B., Laus, M.C., 'Comprehensive overview of the quality of plant- and animal-sourced proteins based on the digestible indispensable amino acid score', *Food Science and Nutrition* 8(10) (2020): https://doi.org/10.1002/fsn3.1809

7. Craddock, J.C., Genoni, A., Strutt, E.F. and Goldman, D.M., 'Limitations with the digestible indispensable amino acid score (DIAAS) with special attention to plant-based diets: a review', *Current Nutrition Reports* 10(1) (2021): https://doi.org/10.1007/s13668-020-00348-8

8. Ibid.

9. Chen, Z., Glisic, M., Song, M., Aliahmad, H.A., Zhang, X., Moumdjian, A.C., Gonzalez-Jaramillo, V., van der Schaft, N., Bramer, W.M., Ikram, M.A., Voortman, T., 'Dietary protein intake and all-cause and cause-specific mortality: results from the Rotterdam Study and a meta-analysis of prospective cohort studies', *European Journal of Epidemiology* 35(5) (2020): https://doi.org/10.1007/s10654-020-00607-6

10. 'Vitamin B12', We Eat Balanced: https://weeatbalanced.com/health-and-nutrition/vitamin-b12/

11. 'Iron', NHS: https://www.nhs.uk/conditions/vitamins-and-minerals/iron/

12. Bastide, N.M., Pierre, F.H.F. and Corpet, D.E., 'Heme iron from meat and risk of colorectal cancer: a meta-analysis and a review of the mechanisms involved', *Cancer Prevention Research* 4(2) (2011): https://doi.org/10.1158/1940-6207.CAPR-10-0113

13. Fang, X., An, P., Wang, H., Wang, X., Shen, X., Li, X., Min, J., Liu, S. and Wang, F., 'Dietary intake of heme iron and risk of cardiovascular disease: a dose-response meta-analysis of prospective cohort studies', *Nutrition, Metabolism and Cardiovascular Diseases* 25(1) (2015): https://doi.org/10.1016/j.numecd.2014.09.002

14.   Tong, T.Y.N., Key, T.J., Gaitskell, K., Green, T.J., Guo, W., Sanders, T.A. and Bradbury, K.E., 'Hematological parameters and prevalence of anemia in white and British Indian vegetarians and nonvegetarians in the UK Biobank', *American Journal of Clinical Nutrition* 110(2) (2019): https://doi.org/10.1093/ajcn/nqz072

15.   Schüpbach, R., Wegmüller, R., Berguerand, C., Bui, M. and Herter-Aeberli, I., 'Micronutrient status and intake in omnivores, vegetarians and vegans in Switzerland', *European Journal of Nutrition* 56(1) (2017): https://doi.org/10.1007/s00394-015-1079-7

16.   Haider, L.M., Schwingshackl, L., Hoffmann, G. and Ekmekcioglu, C., 'The effect of vegetarian diets on iron status in adults: a systematic review and meta-analysis', *Critical Reviews in Food Science and Nutrition* 58(8) (2018): https://doi.org/10.1080/10408398.2016.1259210

17.   Tong, T.Y.N., Appleby, P.N., Armstrong, M.E.G., Fensom, G.K., Knuppel, A., Papier, K., Perez-Cornago, A., Travis, R.C. and Key, T.J., 'Vegetarian and vegan diets and risks of total and site-specific fractures: results from the prospective EPIC-Oxford study' *BMC Medicine* 18(1) (2020): https://doi.org/10.1186/s12916-020-01815-3

18.   Qiao, D., Li, Y., Liu, X., Zhang, X., Qian, X., Zhang, H., Zhang, G. and Wang, C., 'Association of obesity with bone mineral density and osteoporosis in adults: a systematic review and meta-analysis', *Public Health* 180 (2020): https://doi.org/10.1016/j.puhe.2019.11.001

19.   Appleby, P., 'The EPIC-Oxford study and its contribution to the study of vegetarian health', Vegetarian Society: https://vegsoc.org/comment-opinion/epic-oxford-study/

20.   Thorpe, D.L., Beeson, W.L., Knutsen, R., Fraser, G.E. and Knutsen, S.F., 'Dietary patterns and hip fracture in the Adventist Health Study 2: combined vitamin D and calcium supplementation mitigate increased hip fracture risk among vegans', *The American Journal of Clinical Nutrition* 114(2) (2021): https://doi.org/10.1093/ajcn/nqab095

21.   Storhaug, C.L., Fosse, S.K., Fadnes, L.T., 'Country, regional, and global estimates for lactose malabsorption in adults: a systematic review and meta-analysis', *The Lancet Gastroenterology and Hepatology* 2(10) (2017): https://doi.org/10.1016/S2468-1253(17)30154-1

22.   'The vegan diet', NHS: https://www.nhs.uk/live-well/eat-well/how-to-eat-a-balanced-diet/the-vegan-diet/

23.   'Vegetarian, vegan and plant-based diet: food fact sheet', BDA: https://www.bda.uk.com/resource/vegetarian-vegan-plant-based-diet.html

24.   Melina, V., Craig, W. and Levin, S., 'Position of the Academy of Nutrition and Dietetics: vegetarian diets', *Journal of the Academy of Nutrition and Dietetics* 116(12) (2016): https://doi.org/10.1016/j.jand.2016.09.025

25.   'Becoming a vegetarian', Harvard Medical School (15 April 2020): https://www.health.harvard.edu/staying-healthy/becoming-a-vegetarian

26.   'What you need to know about following a vegan eating plan', UnlockFood.ca: https://www.unlockfood.ca/en/Articles/Vegetarian-and-Vegan-Diets/What-You-Need-to-Know-About-Following-a-Vegan-Eati.aspx

27.   'Plant-based diets and their impact on health, sustainability and the environment', World Health Organization: https://apps.who.int/iris/bitstream/handle/10665/349060/WHO-EURO-2021-4007-43766-61591-eng.pdf?sequence=1&isAllowed=y

28.   Deshpande, S. and Singh, R., 'Hemagglutinating activity of lectins in selected varieties of raw and processed dry bean', *Journal of Food Processing*

and Preservation 15(2) (2007): https://doi.org/10.1111/j.1745-4549.1991.tb00156.x

29. Singh, R.S., Kaur, H.P. and Kanwar, J.R., 'Mushroom lectins as promising anticancer substances', *Current Protein and Peptide Science* 17(8) (2016): https://doi.org/10.2174/1389203717666160226144741

30. Liu, Z., Luo, Y., Zhou, T.T. and Zhang, W.Z., 'Could plant lectins become promising anti-tumour drugs for causing autophagic cell death?', *Cell Proliferation* 46(5) (2013): https://doi.org/10.1111/cpr.12054

31. Pusztai, A., 'Dietary lectins are metabolic signals for the gut and modulate immune and hormone functions', *European Journal of Clinical Nutrition* 47(10) (1993): https://pubmed.ncbi.nlm.nih.gov/8269884/

32. Darmadi-Blackberry, I., Wahlqvist, M.L., Kouris-Blazos, A., Steen, B., Lukito, W., Horie. Y, and Horie, K., 'Legumes: the most important dietary predictor of survival in older people of different ethnicities', *Asia Pacific Journal of Clinical Nutrition* 13(2) (2004): https://pubmed.ncbi.nlm.nih.gov/15228991/

33. Ye, E.Q., Chacko, S.A., Chou, E.L., Kugizaki, M. and Liu, S., 'Greater whole-grain intake is associated with lower risk of type 2 diabetes, cardio-vascular disease, and weight gain', *The Journal Nutrition* 142(7) (2012): https://doi.org/ 10.3945/jn.111.155325

34. Gholipour, B., 'No, you probably shouldn't follow Kelly Clarkson's "lectin-free" diet', Live Science (26 June 2018): https://www.livescience.com/62914-what-are-lectins-plant-paradox.html

35. 'Lectin shield', GundryMD.com: https://gundrymd.com/supplements/lectin-shield/

36. Petroski, W. and Minich, D.M., 'Is there such a thing as "anti-nutrients"? A narrative review of perceived problematic plant compounds', *Nutrients* 12(10) (2020): https://doi.org/10.3390/nu12102929

37. Ibid.

38. Borin, J.F., Knight, J., Holmes, R.P., Joshi, S., Goldfarb, D.S. and Loeb, S., 'Plant-based milk alternatives and risk factors for kidney stones and chronic kidney disease', *Journal of Renal Nutrition* 32(3) (2022): https://doi.org/10.1053/j.jrn.2021.03.011

39. Shi, L., Arntfield, S.D. and Nickerson, M., 'Changes in levels of phytic acid, lectins and oxalates during soaking and cooking of Canadian pulses', *Food Research International* 107 (2018): https://doi.org/10.1016/j.foodres.2018.02.056

40. Armah, S.M., Boy, E., Chen, D., Candal, P. and Reddy, M.B., 'Regular consumption of a high-phytate diet reduces the inhibitory effect of phytate on nonheme-iron absorption in women with suboptimal iron stores', *The Journal of Nutrition* 145(8) (2015): https://doi.org/10.3945/jn.114.209957

41. Aranda-Olmedo, I. and Rubio, L.A., 'Dietary legumes, intestinal microbiota, inflammation and colorectal cancer', *Journal of Functional Foods* 64 (2020): https://doi.org/10.1016/j.jff.2019.103707

42. Gonzalez, A.A.L., Grases, F., Mari, B., Tomas-Salva, M. and Rodriguez, A., 'Urinary phytate concentration and risk of fracture determined by the FRAX index in a group of postmenopausal women', *Turkish Journal of Medical Science* 49(2) (2019): https://doi.org/10.3906/sag-1806-117

43. Curhan, G.C., Willett, W.C., Knight, E.L. and Stampfer, M.J., 'Dietary factors and the risk of incident kidney stones in younger women: Nurses' Health Study II', *Archives of Internal Medicine* 164(8) (2004): https://doi.org/10.1001/archinte.164.8.885

44. Sanchis, P., Rivera, R., Berga, F., Fortuny, R., Adrover, M., Costa-Bauza, A., Grases, F. and Masmiquel, L., 'Phytate decreases formation of advanced glycation end-products in patients with type II diabetes: randomized crossover trial', *Scientific Reports* 8(1) (2018): https://doi.org/10.1038/s41598-018-27853-9

45. Petroski and Minich, 'Is there such a thing as "anti-nutrients"? A narrative review of perceived problematic plant compounds'

46. Collins, P., 'Bizarre moment controversial doctor who eats more than a kilo of red meat every day warns that salad is "dangerous" and tells how he hasn't eaten a vegetable in five years', *Daily Mail* (11 June 2023): https://www.dailymail.co.uk/news/article-12182943/Dr-Anthony-Chaffee-warns-60-Minutes-interview-vegetables-trying-kill-you.html

47. Bosetti C. et al., 'Cruciferous vegetables and cancer risk in a network of case-control studies', *Annals of Oncology* 23(8) (2012): https://doi.org/10.1093/annonc/mdr604

48. Su, X. et al, 'Anticancer activity of sulforaphane: the epigenetic mechanisms and the Nrf2 signaling pathway', *Oxidative Medicine and Cellular Longevity* 2018 (2018): https://doi.org/10.1155/2018/5438179

49. Pollock, R.L., 'The effect of green leafy and cruciferous vegetable intake on the incidence of cardiovascular disease: a meta-analysis', *JRSM Cardiovascular Disease* 5 (2016): https://doi.org/10.1177/2048004016661435

50. Salehi B. et al, 'Resveratrol: a double-edged sword in health benefits', *Biomedicines* 6(3) (2018): https://doi.orG/HYPERLINK "https://doi.org/10.3390/biomedicines6030091"10.3390/biomedicines6030091

51. Thirumal, Sivakumar and Balasubramanium, Deepa, 'Phytoalexins: defend systems of plants and pharmacological potential – a systematic review', International Journal of Engineering Technology and Management Sciences 7(2) (2023): https://doi.org/10.46647/ijetms.2023.v07i02.039

52. Wang Y. et al., 'Associations between plant-based dietary patterns and risks of type 2 diabetes, cardiovascular disease, cancer, and mortality – a systematic review and meta-analysis', *Nutrition Journal* 22(1) (2023): https://doi.org/10.1186/s12937-023-00877-2

53. Landsverk, G., 'Some keto evangelists believe vegetable oil is worse than cigarettes, but the science behind the theory doesn't add up', *Business Insider* (6 August 2020): https://www.businessinsider.com/seed-oils-arent-worse-than-cigarettes-despite-keto-claim-2020-7

54. Rett B.S. and Whelan, J., 'Increasing dietary linoleic acid does not increase tissue arachidonic acid content in adults consuming Western-type diets: a systematic review', *Nutrition and Metabolism* 8(36) (2011): https://www.doi.org/10.1186/1743-7075-8-36

55. Johnson, G.H. and Fritsche, K., 'Effect of dietary linoleic acid on markers of inflammation in healthy persons: a systematic review of randomized controlled trials', *Journal of the Academy of Nutrition and Dietetics* 112(7) (2012): https://doi.org/10.1016/j.jand.2012.03.029

56. Su, H., Liu, R., Chang, M., Huang, J. and Wang, X., 'Dietary linoleic acid intake and blood inflammatory markers: a systematic review and meta-analysis of randomized controlled trials', *Food and Function* 8(9) (2017): https://doi.org/10.1039/C7FO00433H

57. Marklund, M. et al., 'Biomarkers of dietary omega-6 fatty acids and incident cardiovascular disease and mortality', *Circulation* 139(21) (2019): https://doi.org/10.1161/CIRCULATIONAHA.118.038908

58. Mousavi, S.M., 'Dietary intake of linoleic acid, its concentrations, and the risk of type 2 diabetes: a systematic review and dose-response meta-analysis of prospective cohort studies', *Diabetes Care* 44(9) (2021): https://doi.org/10.2337/dc21-0438

59. Hajihashemi, P., Feizi, A., Heidari, Z., Haghighatdoost, F., 'Association of omega-6 polyunsaturated fatty acids with blood pressure: a systematic review and meta-analysis of observational studies', Critical Reviews in Food Science and Nutrition 63(14) (2023): https://doi.org/10.1080/104083 98.2021.1973364

60. Li, J., Guasch-Ferré, M., Li, Y., Hu, F.B., 'Dietary intake and biomarkers of linoleic acid and mortality: systematic review and meta-analysis of prospective cohort studies', *American Journal of Clinical Nutrition* 112(1) (2020): https://doi.org/10.1093/ajcn/nqz34

61. Zhang, Y. et al., 'Cooking oil/fat consumption and deaths from cardiometabolic diseases and other causes: prospective analysis of 521,120 individuals', *BMC Medicine* 19(1) (2021): https://doi.org/10.1186/s12916-021-01961-2

62. Amiri, M., Raeisi-Dehkordi, H., Sarrafzadegan, N., Forbes, S.C. and Salehi-Abargouei, A., 'The effects of Canola oil on cardiovascular risk factors: a systematic review and meta-analysis with dose-response analysis of controlled clinical trials', *Nutrition, Metabolism and Cardiovascular Diseases* 30(12) (2020): https://doi.org/10.1016/j.numecd.2020.06.007

63. Hooper, L. et al., 'Reduction in saturated fat intake for cardiovascular disease', *Cochrane Database of Systematic Reviews* 8(8) (2020): https://doi.org/10.1002/14651858.CD011737.pub3

64. 'Ask the expert: concerns about canola oil', Harvard T.H. Chan School of Public Health (13 April 2015): https://nutritionsource.hsph.harvard.edu/2015/04/13/ask-the-expert-concerns-about-canola-oil/

65. Thiagarajan, K., 'Is a vegan diet healthy for children?', BBC Future (1 June 2022): https://www.bbc.com/future/article/20220525-is-a-vegan-diet-healthy-for-kids

66. Desmond, M.A., Sobiecki, J.G., Jaworski, M., Płudowski, P., Antoniewicz, J., Shirley, M.K., Eaton, S., Książyk, J., Cortina-Borja, M., De Stavola, B., Fewtrell, M. and Wells, J.C.K., 'Growth, body composition, and cardiovascular and nutritional risk of 5- to 10-y-old children consuming vegetarian, vegan, or omnivore diets', *The American Journal of Clinical Nutrition* 113(6) (2021): https://doi.org/10.1093/ajcn/nqaa445

67. Ibid.

68. Sutter, D.O. and Bender, N., 'Nutrient status and growth in vegan children', *Nutrition Research* 91 (2021): https://doi.org/10.1016/j.nutres.2021.04.005

69. 'Vegetarian, vegan and plant-based diet: food fact sheet', BDA

70. Melina, Craig, and Levin, 'Position of the Academy of Nutrition and Dietetics: vegetarian diets', *Journal of the Academy of Nutrition and Dietetics* 116(12) (2016): https://doi.org/10.1016/j.jand.2016.09.025

71. American Dietetic Association, Dieticians of Canada, 'Position of the American Dietetic Association and Dietitians of Canada: vegetarian diets', *Journal of the American Dietetic Association* 103(6) (2003): https://doi.org/10.1053/jada.2003.50142

72. *Eat for Health: Australian Dietary Guidelines*, Australian Government National Health and Medical Research Council (2013): https://www.eatforhealth.gov.au/sites/default/files/files/Copyright%20update/n55_australian_dietary_guidelines(1).pdf

73. American Dietetic Association, Dieticians of Canada, 'Position of the American Dietetic Association and Dietitians of Canada: vegetarian diets'

74. Clay, X. 'Why vegan meat substitutes can be worse for your diet than junk food', *Telegraph* (23 December 2022): https://www.telegraph.co.uk/health-fitness/body/vegan-meat-substitutes-bad-diet-healthy-junk-food-2022/

75. Crimarco, A., Springfield, S., Petlura, C., Streaty, T., Cunanan, K., Lee, J., Fielding-Singh, P., Carter, M.M., Topf, M.A., Wastyk, H.C., Sonnenburg, E.D., Sonnenburg, J.L. and Gardner, C.D., 'A randomized crossover trial on the effect of plant-based compared with animal-based meat on trimethylamine-N-oxide and cardiovascular disease risk factors in generally healthy adults: Study With Appetizing Plantfood-Meat Eating Alternative Trial (SWAP-MEAT)', *The American Journal of Clinical Nutrition* 112(5) (2020).

76. Toribio-Mateas, M.A., Bester, A. and Klimenko, N., 'Impact of plant-based meat alternatives on the gut microbiota of consumers: a real-world study', *Foods* 10(9) (2021): https://doi.org/10.3390/foods10092040

77. Alessandrini, R., Brown, M.K., Pombo-Rodrigues, S., Bhageerutty, S., He, F.J. and MacGregor, G.A., 'Nutritional quality of plant-based meat products available in the UK: a cross-sectional survey', *Nutrients* 13(12) (2021): https://doi.org/10.3390/nu13124225

78. 'Food guidelines', Blue Zones: https://www.bluezones.com/recipes/food-guidelines/

79. Heiss, S., Coffino, J.A. and Hormes, J.M., 'Eating and health behaviors in vegans compared to omnivores: dispelling common myths', *Appetite* 118 (2017): https://doi.org/10.1016/j.appet.2017.08.001

80. Melina, Craig, and Levin, 'Position of the Academy of Nutrition and Dietetics: vegetarian diets'

81. Kohl, I.S., Luft, V.C., Patrão, A.L., Molina, M.D.C.B., Nunes, M.A.A. and Schmidt, M.I., 'Association between meatless diet and depressive episodes: a cross-sectional analysis of baseline data from the longitudinal study of adult health (ELSA-Brasil)', *Journal of Affective Disorders* 320 (2017): https://doi.org/10.1016/j.jad.2022.09.059

82. Dobersek, U., Wy, G., Adkins, J., Altmeyer, S., Krout, K., Lavie, C.J. and Archer, E., 'Meat and mental health: a systematic review of meat abstention and depression, anxiety, and related phenomena', *Critical Reviews in Food Science and Nutrition* 61(4) (2020): https://doi.org/10.1080/10408398.2020.1741505

83. Yan, Z., Zeng, X., Su, J, Zhang, X., 'The dark side of empathy: meta-analysis evidence of the relationship between empathy and depression', *PsyCh Journal* 10(5) (2021): https://doi.org/10.1002/pchj.482

84. Holler, S., Cramer, H., Liebscher, D., Jeitler, M., Schumann, D., Murthy, V., Michalsen, A. and Kessler, C.S., 'Differences between omnivores and vegetarians in personality profiles, values, and empathy: a systematic review', *Frontiers in Psychology* 12 (2021): https://doi.org/10.3389/fpsyg.2021.579700

85. Katcher, H.I., Ferdowsian, H.R., Hoover, V.J., Cohen, J.L. and Barnard, N.D., 'A worksite vegan nutrition program is well-accepted and improves health-related quality of life and work productivity', *Annals of Nutrition and Metabolism* 56(4) (2010): https://doi.org/10.1159/000288281

The Practicalist

1. Springmann, M., Clark, M.A., Rayner, M., Scarborough, P., and Webb, P., 'The global and regional costs of healthy and sustainable dietary patterns: a modelling study', *The Lancet Planetary Health* 5(11) (2021): https://doi.org/10.1016/S2542-5196(21)00251-5

2. 'Sustainable eating is CHEAPER as well as healthier' Oxford Martin School (11 November 2021): https://www.oxfordmartin.ox.ac.uk/news/sustainable-eating-is-cheaper/

3. 'Dutch survey finds price gap between meat and meat substitutes is shrinking', ProVeg International (14 April 2022): https://proveg.com/press-release/dutch-survey-finds-price-gap-between-meat-and-meat-substitutes-is-shrinking/

4. Pieper, M., Michalke, A. and Gaugler, T., 'Calculation of external climate costs for food highlights inadequate pricing of animal products', *Nature Communications* 11 (1) (2020): https://doi.org/10.1038/s41467-020-19474-6

5. Springmann, M., Mason-D'Croz, D., Robinson, S., Wiebe, K., Godfray, H.C.J., Rayner, M. and Scarborough, P., 'Health-motivated taxes on red and processed meat: a modelling study on optimal tax levels and associated health impacts', *PLoS ONE* 13(11) (2018): https://doi.org/10.1371/journal.pone.0204139

6. *Growing Better: Ten Critical Transitions to Transform Food and Land Use: The Global Consultation Report of the Food and Land Use Coalition*, Food and Land Use Coalition (2019): https://www.foodandlandusecoalition.org/wp-content/uploads/2019/09/FOLU-GrowingBetter-GlobalReport.pdf

7. *A Multi-Billion-Dollar Opportunity – Repurposing Agricultural Support to Transform Food Systems*, FAO, UNDP and UNEP (2021): https://doi.org/10.4060/cb6562en

ACKNOWLEDGEMENTS

I want to start by stating my appreciation to Penguin Random House for publishing my first two books. It wasn't long ago that the idea of publishing one book was a dream, so to now have two published books is something that I am so very grateful for. In particular, I want to thank Sam Jackson who has presented me with both publishing deals and who has believed in me and the books that I have written. Thank you so much for your trust, your confidence and your desire to have these two books published.

I also want to thank Anya Hayes, who has been instrumental in guiding the book through each stage of the process, and thank you to Jasleen Dhindsa and Shikha Jajoo for their amazing work in marketing, publicity and event organisation. Thank you to Marta Catalano and Alice Brett for proofreading the book so thoroughly and thoughtfully and scrutinising my arguments and citations.

I want to say a huge thank you to Paul Murphy for all of his expertise and help throughout the writing of this book. You helped me better express my arguments and structure my thoughts and I am very grateful for your guidance throughout. A huge thank you also to Hattie Grunewald for her work in securing the book deal.

Getting a book published is a team effort and I am very appreciative of the professionalism, help and work of everyone who has been involved in the publishing of this book.

I want to give a special thank you to my wife Luna for

supporting me throughout the writing and publication of the book, for helping me organise my thoughts and for being a person who inspires me every day.

To everyone who has supported me over the years, thank you for your trust and generosity. Thank you for amplifying my work and for giving me a platform. I am forever thankful to all of you, including those who purchased a copy of my first book *This is Vegan Propaganda (And Other Lies the Meat Industry Tells You)*. Without the support I received for my first book, this book would not exist. So, thank you to each and every one of you.

I also want to acknowledge everyone who is working to create positive change in the world. It's hard and often demoralising to speak out against important issues and it can feel like an uphill battle trying to inspire others to reflect on their behaviours and values. As discussed previously in this book, discussing the ethical, environmental and other issues related to what we do to animals is difficult. Especially considering that often the conversations we end up in are centred around us defending our values from criticisms and scrutiny. Thank you to all of you who have made changes to help protect animals and be vegan, and thank you to all of you who are working to inspire others to do the same.

INDEX

ableism 137–9
acid load, dietary 207
activism 21, 53
agriculture *see* farming
Agriculture and Horticulture
    Development Board
    (AHDB) 253–4
AgriLife 209
air-freight 165
alfalfa 147
algae, red 161
algal oil 259–60, 272, 273
alienation 17
almond milk 88–9, 146–9, 263
alpha-linolenic acid (ALA) 259
Amazon rainforest 150, 222
amino acids 250, 251
anaemia 255–6
ancestors 240–1
Andersen, Inger 199
Anderson, Steven 244
anecdotal evidence, faulty 10–11
animal cruelty 22, 65–7, 79,
    109, 121–3
animal experience 297–9
animal farming 20, 44, 83, 122,
    140, 145, 156, 165–6, 170–90,
    191–5, 199, 204, 205, 214–15,
    222, 230, 233, 245, 291
  and compromise 108
  and cruelty 49
  and deforestation 93, 153
  emissions solutions 160–3, 221
  exposés 65–7
  and global elites 124–7

and insect farming 99, 103
and land use 156–8
and net zero 199–201, 203–4
and racial issues 133, 137
and river pollution 149
and species extinction 51,
    153, 232
stopping with veganism 70–1
and welfare-accredited
    products 113
see also beef; cattle; dairy cattle
animal feed 150–1, 156, 157–63,
    165–6, 194–6, 271–2
additives 124, 160–3, 201
animal rights 53–5
animal welfare 20, 45–52, 111–12,
    173–5
anti-indigenous approach 135–7
anti-nutrients 261–71
antibiotics 91
anxiety 39–40, 61, 68, 76, 246,
    284, 285
apple seeds 266
arachidonic acid 268–9
Archer, Mike 96–7
arguments 44–297
  the amateur nutritionist
    250–86
  the animal farmer 170–90
  the anti-woke warrior 4,
    116–29
  the compromiser 99–115
  the egoist 56–62
  the historian 239–49
  learning your 28–9

arguments – *cont.*
 the misinformed
  environmentalist 146–69
 the mistaken philosopher 45–55
 the naturalist 223–38
 the practicalist 287–96
 the pseudoscientist 191–222
 the social conformer 77–84
 the well-intentioned leftist
  130–45
 the 'what-abouter' 85–98
 the wishful thinker 63–76
Aristotle 117
artificial insemination 170, 171
attention-giving 43
avocado 154–5

bacon 165–6, 207–8, 279
bacteria 91–2
Barclay, Miguel 289–90
Batters, Minette 157, 166
BBC 202
beef 52, 64, 94, 96–7, 100, 147,
  159–63, 164, 166, 197, 209,
  222, 253, 279, 284
bees 86–90, 103–6, 146
beliefs 16–18, 25–6, 29, 45–6
Beston, Henry 297, 298
Beyond Meat 279
biases 10–15, 17–20, 26
Bible 243–8
Biden, Joe 4
Bindel, Julie 167
bioavailability 252, 253, 256
biodiversity 51, 89, 101, 152, 160,
  162, 179, 181, 184–5, 188,
  216, 232
bioenergy 201–4
bioenergy with carbon capture and
  storage (BECCS) 201–2, 204
biofuels 153, 159
biogenic emissions 182–3
biomass, burning 201–2
bivalves 106–8
Black people 132–3

blood sugar spikes 215–18
boats 165
body language 27–8, 38–40, 43
body weight 257, 259
bone mineral density 14, 74–5,
  252, 256–9, 275–6
brain
 development 241–2
 volume 210–12
Brandolini's law 268, 271
Brazil 150–1, 179, 283
British Broadcasting Corporation
  (BBC) 121
British Medical Journal 211, 218
bullshit asymmetry principle *see*
  Brandolini's law

calcium 216, 256–9, 262, 264,
  271, 275–6
California 146–9, 281
calorie restriction 282
cancer 75–6, 204–5, 207–9, 218,
  258, 262, 264–7, 269–70
cannibalism 1, 2, 226
canola oil (rapeseed oil) 269–70
carbon, soil 160, 220–1
carbon capture 188, 189,
  201–2, 220
 see also carbon sequestration
carbon cycle, biogenic 182–3
carbon dioxide 66, 148, 155, 165,
  182–5, 196–9, 202, 215–16,
  219–20
carbon footprint 155, 166
carbon opportunity cost 216
carbon sequestration 148, 155,
  182, 184–5, 201–2, 215–16,
  220–2
carbon stores 152, 160, 162, 183,
  200, 201–2, 204, 216, 221
 see also carbon sequestration
carcinogens 205, 208, 217–18,
  265, 279
cardiovascular disease 207, 218,
  266, 267, 269, 270

carnivores 206
cashew-based products 263, 289
Catholicism 243–4, 248
cats 81, 231–2, 246
cattle 49–50, 81, 96–7, 102, 187, 254
  and 'climate neutral' 200
  and deforestation 151
  and emissions 160–3, 182–3, 197–9
  extinction 51, 52
  grazing 184, 185
  and regenerative farming 159, 222
  seaweed food additives for 124, 160–3, 201
  and water use 147
  wild 51
  zero-grazing systems 161
  see also beef; dairy cattle
Cecil the lion 236–7
chickens 49–51, 61, 63–4, 81, 113, 155, 187, 246, 291
  see also eggs
children 273–7
choice 5–6, 17, 23–4, 70–1, 74, 120, 139, 288
cholesterol 207, 258, 275, 279
choline 211–12
Christianity 130, 243–8
Christmas 131
chronic diseases 74, 133, 207, 209–10, 267, 269, 280, 300
  see also specific diseases
circle of life 228–9
clams 106, 107
clarification 30
classism 139–41
climate anxiety 285–6
climate change 4, 13, 167, 189, 194, 199, 219
climate change deniers 219
climate crisis 20, 24, 137, 145, 221
'climate neutral' 200–1, 203

climate science 24, 166–7, 200, 203
clothing 17, 53, 85, 112, 142
cobalt 254
cocoa 145
Colla, Sheila 106
colonialism 130–2, 135–7
communities 281
companion planting 181
confirmation bias 13–14
conspiracy theories 123, 124–7, 277
consumerism, myth of ethical 141–2
cooking skills 296
coprophagy 254
coronary heart disease 218
Costa Rica 187–8
costs of veganism 132, 289–91
cover crops 180–1
coyotes 234
crickets 100–1, 102
crop residues 194–6
crop rotation 159
crop waste 194–6
crop-deaths argument 94–8
cruciferous vegetables 265–6
cultural appropriation 134–5
culture 20, 80–2, 130–1, 239–40, 242–9
  norms 8–9, 23, 45, 60, 294
cyanide 233, 266

dairy cattle 63–4, 186–7, 201, 215–18, 246, 259
  artificial insemination 170, 171
  and emissions 147–8, 161
  and feed 151, 166
  male calves of 63–4
  and manure 180
  slaughter 63–4, 111–12, 171–2, 298
  water use 147–9
  see also milk, cow's

dairy products 63–4, 99–103, 256, 290
  see also milk, cow's
Davis, Angela 133–4
Davis, Steven 94–5, 97
Dawson, Charles 12, 14
debating skills 27–43
deer 232, 235, 236
deforestation 93, 137, 145, 150–4, 162, 184, 187–8, 193, 220, 222, 245, 300
DEFRA 201
demand 69, 158
dementia 54, 57
Department of Agriculture, Food and the Marine 201
depression 283–6
Derbyshire, Emma 211–12
detachment 19
diabetes 207, 258, 267
  type 2 207, 218, 256, 262, 265, 267, 269
Digestible Indispensable Amino Acid Score (DIAAS) 252–3
dihydrogen monoxide 265
Diodotus 4, 5
disabilities 137–9
discrimination 54
disinformation 8, 12, 211
docosahexaenoic acid (DHA) 259–60
dogs 67–8, 81, 116, 231–3, 246
dolphins 81, 121
Drax power plant 202
dredging 108
ducks 51, 246
Dunning–Kruger effect 15

eating disorders 282–3
eggs 63–4, 99–100, 113, 151, 211–12, 246, 271
  and male chicks 64, 116
ego-centeredness 19, 56–62
eicosapentaenoic acid (EPA) 259–60

Eid 131, 249
Einstein, Albert 298
elk 234, 235
emotionality 22
endorphins 102
ethical issues 16, 52–3, 101–3, 141–2, 235–6
European Union (EU) 137, 153, 157, 164, 185, 290
euthanasia 67–8
exploitation 5, 50, 53, 80, 82, 109–12, 117, 119, 133, 143–5, 149, 175, 233, 248–9, 272
exports 166
extinction 51–2, 153, 162, 230, 234, 245
extremism 79–80, 121–3

fallacy of relative privation 143
families 293–5
far-right 126
farming 3, 13, 18, 20, 21, 44, 49, 125–6, 166–9, 225–7, 231, 237, 241, 246
  and the animal farmer argument 170–90
  and animal welfare 48–50, 113
  arable 145
  and biodiversity 232
  and deforestation 137, 145, 150–4, 162, 187–8, 220, 245
  and egg production 64
  elimination 125
  emissions solutions 160–3
  exposés 65
  and extinction 51–2
  factory 49, 65–6, 144, 291
  fish 71–2, 260
  insect 99–101, 125
  and land use 156–8, 162
  and land yields 94–5, 236
  local 163–7
  and loss of livelihoods 186–90
  and monocropping 158–60, 215
  and predators 233

regenerative 159, 222
and slaughter 66–8
and soya 149–51
stopping 70–1
symbiotic 181
veganic 180–2
and water use 146–9, 154–5
and wildlife 126
see also animal farming
farming unions 189–90
see also National Farmers' Union
fast fashion 53, 85, 112, 142
fats
polyunsaturated 269–70,
275, 276
saturated 218, 258, 268–70,
278–80
see also omega-3 fats; omega-6
fats
fatty liver disease 212
feed conversion ratio 99–100
feelings
fishes' lack of 11, 71–3
of plants 90–3
female genital mutilation 130
Ferrari, Nick 121, 122
fertiliser 158, 159, 178–82
fertility 76, 149
fish 51, 81, 213, 243–4, 248, 260,
271, 281, 298
as not having feelings 11, 71–3
see also pescatarian diets
Fonterra 201
Food and Agriculture
Organization 124
food chains 1, 225, 228–9
food choices 5–6, 17, 23–4, 74,
120, 139, 288
food miles 154, 163–7
food scarcity 239–40, 291–2
food security 157, 179
food waste 52–3, 100–1, 168–9,
194–6
Ford, Henry 186
fortified foods 257, 271–3

fossil fuels 85, 124–5, 166, 183,
187–8, 190, 202, 219
Fox News 4, 17, 210
fractures (bone) 14, 256–9
free-range products 113
friends 295–6

Gates, Bill 123–5, 278
gender differences 285
G.I. Joe fallacy 10
global elites, as pushers of
veganism 123–7
global warming 194, 196–9,
215, 219
global warming potential (GWP)
196–9, 200–1
Glucose Goddess 217–18
glycaemic index 216–17
glycaemic load 217
glycoalkaloids 266
God 244–8
grandfathering 198
grazing 182–3, 184–6, 188–9,
215–18, 219–22
Great Reset 126
green energy 125
green hunting 232
green manure 180–1
greenhouse gas emissions 147–8,
153, 155, 161–2, 164–5, 169,
179–80, 182–3, 185, 191–4,
196–204, 215, 219–22, 290
Greenpeace 202
Guardian (newspaper) 155
Gundry, Steven 261, 262

habitat loss 89, 153, 220, 228, 245
halal 248–9
harm 7, 22, 24, 28, 31–2, 45–9, 53,
57, 59–60, 64, 79–80, 82,
86–7, 90, 92–3, 98, 103, 106,
108–9, 112, 116–17, 119–21,
136, 142, 144–5, 158, 175–7,
225, 228–9, 245, 247–8, 272,
285, 287–8, 299–300

harmful effects of plants 265–7
Harris, Sam 48
Harris, T.I. 277–8
health 14, 26, 83–4, 133, 204–12,
    250–86
  long-term 280–1
  risks of animal products to
    126–7
heart attack 267
heart disease 207–10, 212, 255,
    258, 262, 270
herbivores 236
hexane 270
homosexuality 78, 81, 130
honey 103–6
honeybees 87–8, 103–6
honour killings 130
Humane Research Council 82
humane-washing 66–9, 113
hunter-gatherers 230–1
hunting 230–7

imports 165–6
Impossible Burger 277–8
impressions, good 41–2
inconvenience of veganism 287–8
inflammation 268–9
insect farming 99–103, 125
instrumental insemination 104
intelligence 57–9
Intergovernmental Panel on
    Climate Change (IPCC)
    124, 126, 183
International Life Science Institute
    (ILSI) 209
iron, dietary 255–6, 264, 271
isoflavones 75
Israel 154

JBS 222
Jesus 131, 246–7
Jews 247, 249
Johnson, Boris 203
Johnston, Bradley 208–9
Jones, Alex 277

Kaporos 249
Kennedy, Martin 157, 165, 174–5
kidney stones 262–3, 265
King, Martin Luther, Jr. 301
kosher 248–9

lab-grown meat 229–30
labour, forced 53, 144
Labour government (Wales) 126
lamb 64, 81, 102, 164, 166,
    172–3, 253
land use 156–8, 162, 164,
    184–6, 193, 203, 215–16,
    219–20, 290
land yields 94–5, 236
language use 35–40, 60–2, 213–15
laws, vegan 121–3
leaky gut syndrome 261
lectins 261–2
legality 78–9, 81
legumes 150, 181, 193, 211,
    218, 252, 255, 257, 261–2,
    264, 281
lettuce 165
life cycle of food 164
life expectancy, blue zones 280–1
linoleic acid 268–9
listening skills 29–31, 32–3
local farming 163–7

M-44 cyanide bombs 233
maize 159, 166
malnutrition 274
man-boobs 74–6, 149
manure 178–82
masculinity, symbols of 129
mealworms 100
meat see processed meat; red meat;
    specific meats
Meat & Livestock Australia
    (MLA) 200
meat-free alternatives 73–4,
    213–15, 277–80, 287,
    289–90
Melville, James 125

menopause 75
mental health 283–6
methane 124, 161–3, 181, 182–3, 185, 196–9
methane masks 162
Mexico 154
mice 95, 96
microbes 91–2
microbiome 279
milk
    cow's 63–4, 100, 151, 211, 213, 259
    dairy 215–18
    plant-based 75, 88–9, 146–50, 166, 195, 213–18, 253, 257, 263, 274
minorities 54
mirroring 40
misinformation 8, 10, 12, 14, 26, 126
mobility levels 281
monocropping 158–60, 215
moral issues 9, 23–5, 45–7, 78, 239–41, 246–7
moral responsibility 81
Morgan, Piers 4, 117, 146
motivated reasoning 13–14
muscle mass 238, 251, 252
Muslims 131, 247, 249
mussels 106–8

National Farmers' Union (NFU) 148, 149, 157, 166–7, 201, 202–3
    Scotland 157, 174
National Pig Association (NPA) 113
natural foods 229–30
nature 185, 225–8, 232–5
Nature Scot 159
Neanderthals 12, 240
net zero, meat company pledges towards 199–204
net-positive theory 48–9, 50
Nirvana fallacy 97–8

nitrogen 178–9, 181
nitrogen fixation 181
nitrogen pollution 162
No Farmers, No Food 125
nociceptors 102
Nugent, Ted 94
nutrient deficiencies 250–60, 271–7, 286
nutritional supplements 271–3, 275–6

oat milk 89, 213, 215–18, 263
omega-3 fats 259–60, 271, 273
omega-6 fats 268
Oonincx, Dennis 100
organic produce 159–63, 180, 279, 280, 289
oxalates 262–3
oysters 106–8

pain 102, 104
    inability to feel 11, 71–3
palm oil 152–3
Palmer, Walter 237
Panorama (TV show) 202
Payments for Environmental Services (PES) programme 187–8
pea milk 216
personal freedom 121–3
persuasion 27–8
Peru 167–8
pescatarian diets 283
pesticides 158, 159
Peterson, Jordan 53–4, 125, 206
Phillippi, Dan 67–8
Phillippi, Nikki 67–8
photosynthesis 185, 199
physical exercise, weight-bearing 257, 259
physical 'weakness', and meat deprivation 123–4
phytates (phytic acid) 264–5
phytoalexins 266–7
phytoestrogens 74–6

pigs 47, 49, 51, 61, 66, 81, 246
  and feed 151
  and the gas chamber 47, 66, 93,
    110–11, 116, 174–5, 280
  and gestation crates 61, 113,
    173–4
  see also pork
Piltdown Man 12–13, 14
pleasure-seeking 19–20, 25, 56–7
polarisation 7
political identities 17–20, 26
pollination 86–90
pollinators 86–90, 103–6, 146, 188
pollution 149, 180, 198
pork 151, 155, 160, 166,
    173–4, 253
potatoes 266
poultry 151, 195–6
  see also chickens; ducks; turkeys
poverty 132, 291–2
power, symbols of 127–9
predators 233–5, 236–7
presentation skills 27–8
privilege 139–41
processed meat 205, 207–8,
    217–18, 279, 290
protein 99–100, 102, 250–3,
    258–9
Pythagorean diet 242

QI (TV show) 86
questions, asking 31–5, 43
quinoa 167–8, 274

racism 131–5
rainforests 15, 149–51, 222
randomised controlled trials 205,
    206–7, 284
rational thinking 18
raw-food diets 252, 259, 263,
    264, 274
recidivism 82–4
red meat 4, 14, 99, 160, 200, 203,
    205–10, 280
  life cycle 164

  see also beef; lamb; pork; veal
reforestation 200
religion 131, 243–9
repetition 22–3
Republicans 17
respect 114–15, 120
resveratrol 267
rewilding 162, 185, 189, 199,
    215–16, 291
rights 53–5, 145
Ritchie, Hannah 163
Rogan, Joe 82, 84, 206
Rogelj, Joeri 198
Roundup 209
Royal Society for the Prevention
    of Cruelty to Animals
    (RSPCA) 111

sacrifice 131, 249
Savory, Allan 220–1
Schofield, Philip 207–8
science, 'weak' 204–7
scientific basis 14–15
Sears, J.P. 123, 125
seaweed, as food additives for
    cattle 124, 160–3
seed oils 268–71
self-awareness 12
selfishness 19–20
sentience 22, 59, 61–2, 73, 91–3,
    101–2, 104, 107–8
sharks 121, 237
sheep 49, 51, 185, 233, 236–7,
    246, 254
shellfish 69, 106–8
silage 159, 271
slaughter 49–50, 65–7, 79,
    177–8, 194
  and animal welfare 111–12,
    174–5
  of chickens 64, 113, 291
  of dairy cattle 63–4, 111–12,
    171–2, 298
  ethical 248–9
  'humane' 67–9

inflicted by vegans 94–8
insects 101
of pigs 47, 66, 93, 110–11, 116,
    174–5, 280
see also throat-cutting
slaughterhouses 49, 58, 68, 96,
    115, 171, 215, 246, 298, 300
    workers 70, 143–4
slavery 46, 54–5, 81
slurry 159, 180
smoking 186, 204–5, 209
social contract theory 53–5
social control 123
social media 10, 14–15, 206
Socrates 29, 31, 117
Socratic method/questioning 31–4
soil
    carbon 160, 220–1
    erosion 159, 180–1
    health 100, 158–9, 180–1, 193,
        254, 272
solanine 266
South America 137, 147, 149–51,
    154, 167–8, 187–8
soya 15, 74–6, 100, 137, 149–51,
    165–6, 195–6
soya milk 75, 150, 166, 195, 216,
    253, 257, 263
soya protein formulas (SPFs) 75
soybean meal 194
soybean oil 194
sperm counts 76
Stanley, Joe 166
status symbols 127–9
stereotypes 21–2
stroke 207, 270
stunning 68
subjugation 123, 126, 127
subsidies 49, 122, 140, 177,
    179–80, 291
suffering 5, 19–20, 22, 26, 28,
    48–50, 53–4, 56–62, 64, 66–9,
    71–4, 77–82, 95–6, 101, 103,
    108–10, 113, 116–17, 119,
    131, 133, 142–5, 176–7, 227,

229, 231, 235, 245–7, 249,
    272, 286, 291, 299–300
sugar 208–9
suicide, male 285
sulforaphane 266
Sunak, Rishi 126
'superiority', human 57–9
supply 69
sustainability 6, 16, 24, 49, 78, 80,
    85, 90, 99–101, 103, 109, 121,
    145, 153, 160, 163, 165–6,
    179, 202, 227, 230–6, 272,
    288, 300

technology 237–8
    theoretical 162
testosterone 74–5
Texas A&M 127–8
throat-cutting 32, 50, 56, 66–8, 79,
    174–5, 246, 248, 298
time issues 288
tobacco 186, 204–5, 219
tofu 278
Toksvig, Sandi 86
tomatoes 165
tradition 130–1, 239–40, 242–9
transporting food 163–7
tribalism 2, 17
trimethylamine-N-oxide (TMAO)
    212, 279
turkeys 51, 131, 246
Tutu, Desmond 42
Tyson, Neil deGrasse 91, 299

United Nations (UN) 94, 124, 166
    see also Intergovernmental Panel
        on Climate Change
uric acid 263
UV rays 268

validation 36–7, 43
Vatican 244
veal 64
V.E.G.A.N. acronym 42–3
vegetables 265–6

vitamin B12 210–12, 253–4,
    258–9, 261, 273, 275–6
vitamin D 258–9, 271, 273, 275–6
voice, quality of 38

water pollution 164
water use 146–9, 154–5, 164, 216
whales 81, 298
wheat 95–6
wholegrains 211, 218, 255, 261–2,
    264, 281, 289
wild animals 176–7, 223–4
Wildlife Services 233

Willett, Walter 209
Willoughby, Holly 207–8
woke 4, 116–29, 131–2
wolves 234–5
wood pellets, burning 202
World Economic Forum (WEF)
    123, 125
World Health Organization (WHO)
    126, 205, 208, 209, 260

X 125

Yellowstone National Park 234–5

ABOUT THE AUTHOR

On 14 May 2014, Ed Winters was reading the news online when the headline *Hundreds of chickens killed in M62 lorry crash* grabbed his attention. He clicked on the article and was stunned to read that about 1,500 chickens had been killed in a truck that had crashed while on the way to a slaughterhouse. This article became a huge catalyst in Ed's life, first encouraging him to go vegetarian before he eventually became vegan in 2015.

In early 2016, Ed set up his YouTube channel, where he began uploading street interviews with members of the public on the ethics of eating animals. In the same year, he also co-founded the animal rights organisation Surge, as well as The Official Animal Rights March, a global event which grew from 2,500 participants in London in 2016 to 41,000 participants across the world in 2019. Ed continues to be the co-director of Surge, with the organisation focusing on producing films that tell the untold stories of animals, as well as educational programmes for universities.

Just over a year after co-founding Surge, he released the documentary *Land of Hope and Glory*, which is an exposé of UK terrestrial animal farming. Ed then began touring UK universities to screen the documentary to students, where he also gave a speech to the students after the screening.

His university speech *You Will Never Look at Your Life in the Same Way Again*, which has been given to thousands of students across UK universities, went viral in early 2018 and to date has 35 million accumulative views online.

Regularly lecturing on animal ethics and the environment, Ed has spoken at over one-third of UK universities and at every Ivy League college, including as a guest lecturer at Harvard University. He has given speeches across the world, including at LinkedIn, American Express, Pinterest, Google and Meta. Ed has additionally given two TEDx talks, surpassing a total of 2.5 million views online.

Ed has debated numerous times on live television and radio, and has been featured on the BBC, ITV, LadBible and GB News, and he has been featured by print and online news organisations including The *Guardian*, the *Independent* and *The Times*. In December 2020, Ed announced that he and his team had founded Surge Sanctuary, a forever home for abused and unwanted animals on an 18 acre site in rural England.

In 2021, Ed wrote his debut book *This is Vegan Propaganda (And Other Lies the Meat Industry Tells You)*, which was published by Penguin Random House on 6 January 2022, becoming an instant bestseller. In 2022, Ed became a Media & Design Fellow at Harvard University, teaching a module during the fall semester. In 2023, he won a Webby People's Voice Award for his short film *Milk*.